Darwin

Past Masters

AQUINAS Anthony Kenny
FRANCIS BACON Anthony Quinton
BERKELEY J. O. Urmson
BURKE C. B. Macpherson
CARLYLE A. L. Le Quesne
COLERIDGE Richard Holmes
CONFUCIUS Raymond Dawson
DANTE George Holmes
DARWIN Jonathan Howard
ENGELS Terrell Carver

GALILEO Stillman Drake
HOMER Jasper Griffin
HUME A. J. Ayer
JESUS Humphrey Carpenter
KANT Roger Scruton
MACHIAVELLI Quentin Skinner
MARX Peter Singer
MONTAIGNE Peter Burke
PASCAL Alban Krailsheimer
TOLSTOY Henry Gifford

Forthcoming

ARISTOTLE Jonathan Barnes
AUGUSTINE Henry Chadwick
BACH Denis Arnold
BAYLE Elisabeth Labrousse
BERGSON Leszek Kolakowski
THE BUDDHA Michael Carrithers
JOSEPH BUTLER R. G. Frey
CERVANTES P. E. Russell
CHAUCER George Kane
CLAUSEWITZ Michael Howard
COBBETT Raymond Williams
COPERNICUS Owen Gingerich
DIDEROT Peter France
ERASMUS James McConica
GIBBON J. W. Burrow
GODWIN Alan Ryan
GOETHE T. J. Reed
HEGEL Peter Singer
HERZEN Aileen Kelly
JEFFERSON Jack P. Greene
JOHNSON Pat Rogers

LAMARCK L. J. Jordanova
LINNAEUS W. T. Stearn
LOCKE John Dunn
MENDEL Vitezslav Orel
MILL William Thomas
THOMAS MORE Anthony Kenny
MORRIS Peter Stansky
MUHAMMAD Michael Cook
NEWMAN Owen Chadwick
NEWTON P. M. Rattansi
PETRARCH Nicholas Mann
PLATO R. M. Hare
PROUST Derwent May
RUSKIN George P. Landow
ST PAUL Tom Mills
SHAKESPEARE Germaine Greer
ADAM SMITH A. W. Coats
SOCRATES Bernard Williams

and others

Jonathan Howard

DARWIN

Oxford Toronto Melbourne
OXFORD UNIVERSITY PRESS
1982

Oxford University Press, Walton Street, Oxford OX2 6DP

London Glasgow New York Toronto
Delhi Bombay Calcutta Madras Karachi
Kuala Lumpur Singapore Hong Kong Tokyo
Nairobi Dar es Salaam Cape Town
Melbourne Auckland

and associate companies in
Beirut Berlin Ibadan Mexico City

First published as an Oxford University Press paperback
1982 and simultaneously in a hardback edition

British Library Cataloguing in Publication Data

Howard, Jonathan
Darwin. – (Past masters)
1. Darwin, Charles – Influence
2. Evolution – History
3. Intellectual life – History
I. Title II. Series
909.8   QH31.D2
ISBN 0-19-287557-4
ISBN 0-19-287556-6 Pbk

Printed in Great Britain by
Cox & Wyman Ltd, Reading

Preface

With the centenary of Darwin's death comes a widespread mood of scepticism and unease about the validity and significance of Darwin's contribution to knowledge. The time is certainly right to try to describe Darwin's scientific work briefly and in plain language. This book is not unbiased, in the sense that I do not stand equidistant between Darwin and his detractors. No working biologist can read and understand Darwin's work without realising the overwhelming importance it has had for the development of biological thought in general. But I hope the book is fair in pointing out aspects of Darwin's thinking which lack consistency or have failed to stand up to critical scrutiny. Perhaps to emphasise the importance of Darwin for biology will be considered as limiting the scope and reference of Darwinism. But that would be nonsense. The whole point of understanding the modern theory of evolution is to understand that human life and human society are to a certain extent biological issues, painfully difficult to deal with in these terms, but still biology. For this reason I have kept strictly to the core of the matter, which is Darwin's contribution to biology. Darwinian philosophy or Darwinian society are post-hoc constructs that had no place in Darwin's thought. He was unable to see how evolution in biology could have any but the feeblest analogical resemblance to the evolution of society. The complete separation between social and political philosophy and Darwin's Darwinism is the main justification, if any is needed, for dealing only with the latter. The biology is where the issues are finally grounded, and it is probably the biology that is least generally known.

My Darwinising has imposed heavily on the kindness and forebearance of my friends and close scientific colleagues, and I apologise and thank them all. I owe a particular debt of thanks to Geoffrey Butcher who took over the burden of running the laboratory during my prolonged absences.

Thanks also to Henry Hardy at the Oxford University Press

for the invitation to write this book, and to Michael Singer for a chance remark which tilted the balance in favour of accepting the invitation. I am exceedingly grateful to Keith Thomas, Henry Hardy and Virginia Llewellyn Smith for their invaluable critical comments on the text, and for their editorial help in cutting this particular Past Master down to size.

Finally I should like to thank Mrs Susan Smith and Mrs Christine Strachan for their quick and intelligent typing of a difficult manuscript.

JONATHAN HOWARD

Abbreviations

The following are the works by Darwin from which I have quoted in the text. References to Darwin's writings are given by an abbreviation of the title, followed by volume number where appropriate, and a page reference.

A   *Charles Darwin, Thomas Henry Huxley. Autobiographies*, edited with an introduction by Gavin de Beer, Oxford, 1974.

D   *The Descent of Man and Selection in Relation to Sex*, 2 vols, London, 1871.

F   *The Various Contrivances by which Orchids are Fertilized by Insects*, London, 1890.

J   *Journal of Researches into the Natural History and Geology of the Countries Visited during the Voyage of H.M.S. 'Beagle' round the World*, London, 1882.

L   *Life and Letters of Charles Darwin*, edited by F. Darwin, 3 vols, London, 1888.

M   *'Darwin's Early and Unpublished Notebooks', transcribed and annotated by Paul H. Barrett, in Howard E. Gruber, *Darwin on Man; A Psychological Study of Scientific Creativity*, London, 1974.

ML   *More Letters of Charles Darwin*, edited by F. Darwin and A. C. Seward, 2 vols, London, 1903.

O   *On the Origin of Species. A facsimile of the First Edition*, with an introduction by Ernst Mayr. Fourth Printing. Cambridge, Mass., 1976

T   *Darwin's Notebooks on the Transmutation of Species*, edited by Gavin de Beer and others. Bulletin of the

* Still in print

British Museum (Natural History). Historical Series. Parts I, II, III, IV and VI. London, 1960 and 1967.

V   *The Variations of Animals and Plants under Domestication*, 2 vols, London, 1868.

Contents

To my mentors
J. L. Gowans and
P. B. Medawar

1 Darwin's life

Darwin's biographers are faced with an embarrassment of riches. His parents were both children of distinguished families that have earned biographical attention in their own right. Darwin then married his first cousin, and the family seems to have thrown practically nothing away ever since. Darwin recorded his own life in an autobiography written privately for his family, but naturally kept, and since published. The notes and records of a whole lifetime's scientific work have been maintained virtually intact. Darwin lived in only three places during a scientific career lasting fifty years; five years aboard the *Beagle* while circumnavigating the world, four years in London, and the remainder in Down House, a few miles south of London. The theory of evolution started on the *Beagle* voyage. His library on board is known, his notes and journals from the voyage survive. Darwin wrote one long book about the voyage, while the captain of the *Beagle* wrote another. Many of the specimens that Darwin collected are still together. There is even a pictorial record of the *Beagle* voyage from the two artists that sailed with the ship. Returning home, Darwin wrote a series of personal notebooks which record the earliest developments of the theory of evolution in the most idiosyncratic and fascinating detail. Two complete preliminary versions of the theory of evolution survive, one brief and in pencil, the other longer and carefully transcribed in ink.

Ill for the last forty-five years of his life, Darwin worked through an enormous correspondence. Five volumes of his letters were edited by his son Francis, and a few more letters have appeared sporadically since. A definitive edition of more than 13,000 letters is planned. Many of Darwin's correspondents were scientists of great distinction. Their own correspondence with Darwin in turn survives from an age when a 'Life and Letters' was the conventional celebration of a great man's passing. Finally, Darwin prepared an immense amount of scientific material for publication, from short notes and questionnaires through longer papers to a succession of major books. His complete published work is catalogued in Freeman's excellent bibliography (see Further reading).

Had Darwin been a lesser figure, such a stupendous collection of biographical sources would still guarantee him a place in the history of nineteenth-century science. As it is, there is an almost unbelievably complete documentary record of the life of one of the great revolutionaries in the history of ideas. Small wonder, then, that the exploration of this record and its implications for the history of science has become what is sometimes referred to as the Darwin Industry.

Charles Darwin was born in Shrewsbury in 1809. His father was a doctor, son of a more distinguished doctor and speculative evolutionist, Erasmus Darwin. His mother was a daughter of Josiah Wedgwood, founder of the pottery. She died when Charles was eight, leaving him to be brought up largely by his sisters. He was educated at the local public school and went to Edinburgh University as a medical student. Unable to face confrontation with the seriously ill, Darwin abandoned medicine and moved from Edinburgh to Cambridge with the intention of becoming an Anglican priest. While at Cambridge he became the protégé of the Professor of Botany, John Stevens Henslow, through whom he became deeply interested in science, and through whose advocacy Darwin was chosen at the age of twenty-two to go on a surveying voyage with HMS *Beagle* as naturalist. The *Beagle*'s circumnavigation of the globe lasted five years, from 1831 to 1836. In his autobiography Darwin summed up the impact of the *Beagle* voyage on his life.

The voyage of the Beagle has been by far the most important event in my life and has determined my whole career . . . I have always felt that I owe to the Voyage the first real training or education of my mind. I was led to attend closely to several branches of natural history, and thus my powers of observation were improved, though they were already fairly developed. The investigation of the geology of all the places visited was far more important, as reasoning here comes into play. On first examining a new district nothing can appear more hopeless than the chaos of rocks; but by recording the stratification and nature of the rocks and fossils at many points, always reasoning and predicting what will be found elsewhere, light soon begins to dawn on the district, and the structure of the whole becomes more or less intelligible. I had brought with me the first volume of Lyell's Principles of Geology, which I studied attentively; and this book was of the highest service to me in many ways. The very first place which I examined, namely St. Jago in the Cape Verde islands, showed me clearly the wonderful superiority of Lyell's manner of treating geology, compared with that of any other author, whose works I had with me or ever afterwards read. Another of my occupations was collecting animals of all classes, . . . but from not being able to draw, and from not having sufficient anatomical knowledge a great pile of M.S which I made during the voyage has proved almost useless . . . During some part of the day I wrote my Journal, and took much pains in describing carefully and vividly all that I had seen; and this was good practice . . . The above various special studies were, however, of no importance compared with the habit of energetic industry and of concentrated attention to whatever I was engaged in, which I then acquired. Everything about which I thought or read was made to bear directly on what I had seen or was likely to see; and this habit of mind was continued during the five years of the voyage. I feel sure that it was this training which has enabled me to do whatever I have done in science. . . .

As far as I can judge of myself, I worked to the utmost during the voyage from the mere pleasure of investigation, and from my strong desire to add a few facts to the great mass of facts in natural science. But I was also ambitious to take a fair place amongst scientific men, – whether more ambitious or less so than most of my fellow-workers I can form no opinion. (A 44–6)

Darwin returned from the *Beagle* voyage as a promising geologist. Two discoveries, a brilliant integrative theory of the origin and distribution of coral reefs, and a more conventional but still impressive account of the rapid land elevation continuing to form the Andean mountain chain, brought him the respect of the greatest geologist of the age, Charles Lyell. It was the beginning of a lifetime's friendship. Lyell was the outstanding exponent of scientific evolutionary geology, a methodological innovation in the study of earth history which urged that presently acting and knowable geological processes rather than miraculous interpositions of divine power were sufficient to account for the evolution of the earth's crust. The second volume of Lyell's *Principles of Geology*, dealing with biological evolution and the distribution of animals and plants, reached Darwin while the *Beagle* was in South America. Although Lyell rejected biological evolution, the *Principles of Geology* was the only work of scientific distinction that Darwin read on the *Beagle*, indeed one of the very few such books that he had ever read, and, as emphasised in the next chapter, its influence on the scientific outcome of the *Beagle* voyage was overwhelming.

> I always feel as if my books came half out of Lyell's brain, and that I never acknowledge this sufficiently; . . . for I have always thought that the great merit of the Principles was that it altered the whole tone of one's mind. (ML i. 177)

Observations on the relationship between living and fossil mammals in South America, and the South American appearance of unique plant and animal species found on the Galapagos Archipelago off the coast of Ecuador, eventually convinced Darwin that Lyell was wrong about biological evolution.

In the three years from 1837 to 1839, Darwin produced the complete theory of evolution in about 900 pages of private notes written in his spare time. All the main issues which occupied him for the rest of his life were dealt with piecemeal in a torrent of creative insight of extraordinary intensity. Darwin developed his evolutionary ideas on many fronts with great rapidity, and the Notebooks do not convey an ordered process of accumulation and rationalization. The theory was first recorded as a consecutive argument in a thirty-five-page 'sketch' written in 1842, followed by a 230-page Essay in 1844. Neither of these productions was intended for publication, but Darwin made elaborate arrangements with his wife for the publication of the 1844 essay in the event of an early death.

In 1839 Darwin published the first version of the *Journal of*

Researches describing the events and discoveries of the *Beagle* voyage for a popular audience. To his surprise, this quickly became one of the most widely read travel books of the nineteenth century. Little hint of the evolutionary ideas appeared in the *Journal of Researches*, even after a new edition was prepared in 1845. From 1842 to 1846 Darwin published the geological results of the *Beagle* voyage in three volumes as part of the official record of the expedition.

In 1839 Darwin married Emma Wedgwood, and their first child was born in 1841. Darwin's robust health had begun to collapse and in 1842 the family left London for Down House in Kent. Apart from occasional trips to London or to visit relatives, Darwin left Down House thereafter only to visit health spas where he subjected himself uncritically to nonsensical and unpleasant hydropathic 'cures'. Prolonged researches have not established what was wrong with Darwin. He was easily exhausted, insomniac and had recurrent intestinal pains and nausea. Throughout the rest of his life he worked around his illness, a few hours a day if possible, making careful notes about the number of days lost.

Darwin's isolation in Down House was relieved by close friendship with the best scientists of the time, especially Lyell and the botanist Joseph Hooker, who alone had the privilege of reading the 1844 essay shortly after it was written. While continuing to accumulate information on the species question, Darwin began a wholly new piece of research, the classification of a little understood group of marine crustacea called barnacles. This immense study was published between 1851 and 1854 and immediately became the definitive text on the subject. Darwin felt the enterprise had not justified the eight years' labour, and his judgement was probably correct. Nevertheless, classification and the definition of species was a field that Darwin wished to revolutionise with the theory of evolution, and the work was certainly begun with a view to discovering the problems of classification at first hand. Despite the importance of the theory of evolution to Darwin's whole approach to the work, no explicit reference to the theory appears in the barnacles books.

By 1854 Darwin felt ready to present the whole case for the theory of evolution in public, and, urged by his friends, began to organise all the material collected over seventeen years into a colossal book. In the event this was never completed because in 1858 he received from a correspondent in Borneo, Alfred Russel Wallace, a letter enclosing a short essay which outlined in the most succinct terms the whole of Darwin's theory. Embarrassed and piqued, Darwin did not know what to do, and passed the essay to Lyell for his opinion. Lyell and Hooker persuaded Darwin to allow some of his own material, an extract from the 1844 essay and part of a letter of 1857 to the American botanist Asa Gray, to be read with Wallace's essay to the Linnean Society in London. So little notice was taken of this event that the then President of the Linnean Society felt able to summarise the year's proceedings, perhaps the most momentous proceedings of any learned society at any time, in the following terms: 'The year . . . has not, indeed, been marked by any of those striking discoveries which at once revolutionize, so to speak, the department of science on which they bear.'

Provoked, anxious for the recognition of his twenty-year priority, ill, and grieving over the death of one of his beloved children, Darwin set aside the enormous book and began to write an 'abstract' of the work. This was eventually published at the end of 1859 as *The Origin of Species by Means of Natural Selection, or the Preservation of Favoured Races in the Struggle for Life*.

The *Origin* produced an instant reaction from the public and from the scientific community. An intemperate battle began in newspapers, magazines and scientific assemblies. Darwin's most fervent supporter was the young Thomas Henry Huxley, a brilliant and eloquent anatomist, who fought for the *Origin* while Darwin hid in Down House tinkering with the text of successive editions. At the meeting of the British Association for the Advancement of Science in Oxford in 1860, the Bishop of Oxford, Samuel Wilberforce, determined to 'smash Darwin' in a session chaired by Henslow and attended by Hooker and Huxley. Unfamiliar with his subject, Wilberforce

spoke for full half-an-hour with inimitable spirit, emptiness and unfairness . . . Unfortunately the Bishop, hurried along on the current of his own eloquence, so far forgot himself as to push his attempted advantage to the verge of personality in a telling passage in which he turned round and addressed Huxley: I forget the precise words, and quote from Lyell. 'The Bishop asked whether Huxley was related by his grandfather's or grandmother's side to an ape.' (L ii. 321–2)

Huxley turned to his neighbour, saying 'The Lord hath delivered him into mine hands' and, after dealing briskly with the few scientific matters raised by Wilberforce, won the day for Darwin with a crushing rebuke:

I asserted . . . that a man has no reason to be ashamed of having an ape for his grandfather. If there were an ancestor whom I should feel shame in recalling, it would be a *man*, a man of restless and versatile intellect, who, not content with an equivocal success in his own sphere of activity, plunges into scientific questions with which he has no real acquaintance, only to obscure them by an aimless rhetoric, and distract the attention of his hearers from the real point at issue by eloquent digressions, and skilled appeals to religious prejudice. (L ii. 322)

The British Association was an important forum in the midnineteenth century, and as Darwin justly wrote to Huxley on hearing of the Oxford fracas,

From all that I hear from several quarters, it seems that Oxford did the subject great good. It is of enormous importance, the showing the world that a few first-rate men are not afraid of expressing their opinion. (L ii. 324)

He also wrote, no doubt with equal truth, 'I honour your pluck; I would as soon have died as tried to answer the Bishop in such an assembly . . .' (L ii. 324).

From 1860 until his death Darwin produced a succession of major works developing various themes in evolution. Most of these themes had been at least touched on in the *Origin*, but each new book was strikingly original both in approach and subject matter. Two works on man and a book on variation under domestication developed trains of thought that Darwin had sketched out in the great years 1837–9. Three books on the sexual biology of flowers, two books on climbing plants and other aspects of plant movement, and one book on insectivorous plants reflected Darwin's interest in the general problem of the nature of biological adaptation combined with his ability to experiment only in his own gardens and greenhouses because of the limitations imposed by ill-health. His last book, published in 1881, was on the role of earthworms in the formation of vegetable mould, in which he returned to a subject that he had initiated in his first scientific publication in 1838.

Darwin's illness and a natural diffidence kept him from public life. He gave no public lectures on evolution and wrote nothing on the subject more accessible than the *Origin of Species*. His fame grew as much from the enthusiasm of his admirers as from his own efforts. T. H. Huxley's tenacious defence of Darwin began with an almost Pauline revelation when he first learned of the principle of natural selection. 'How extremely stupid not to have thought of that,' he later remembered thinking. From 1859 until his death in 1894, Huxley rammed the Darwinian argument down the most unreceptive throats he could find. Combative and clever, he forced the conflict between science and scripture out into the open and demolished any attempt at reconciliation. Through popular lectures to working men he brought evolution to ordinary people for whom elevated sectarian conflicts were an irrelevancy. It is very much to Huxley's credit that the Darwinian revolution was so rapid, and that it was seen as a revolution by all who lived through it.

Darwin's pre-eminent contribution to science was never formally recognised during his lifetime by the Royal Society. When he was awarded its highest honour, the Copley Medal, in 1864, the citation was expressly formulated to exclude the theory of evolution. The Royal Society made amends after his death by founding a Darwin Medal whose first three recipients were consecutively Wallace, Hooker and Huxley.

Darwin inherited enough money from his family to ensure that he never needed to work for a living. He added to this with income from the relatively enormous sales of his books, and managed his financial affairs skilfully. Towards the end of his life he became very rich. He was revered by his intimate scientific friends and loved by his family, to whom he was devoted.

The family lost three children, each bereavement causing Darwin the most intense grief. Cruelty in any form roused him to passionate anger: there was no passage in the *Journal of Researches* more deeply felt than where he attacked the institution of slavery. This was the only issue over which he quarrelled with Lyell.

In religion Darwin moved from a conventional orthodoxy in youth to a sceptical agnosticism which he never lost:

> disbelief crept over me at a very slow rate, but was at last complete. The rate was so slow that I felt no distress, and have never since doubted even for a single second that my conclusion was correct. I can indeed hardly see how anyone ought to wish Christianity to be true; for if so, the plain language of the text seems to show that the men who do not believe, and this would include my Father, Brother and almost all my best friends, will be everlastingly punished.
>
> And this is a damnable doctrine. (A 50)

Darwin reserved this uncompromising expression of his religious views for the private autobiography that he wrote for his family. Had he not done so it is inconceivable that T. H. Huxley could have abandoned his war against hypocrisy and helped to carry Darwin's coffin into Westminster Abbey in 1882.

2 The foundations of Darwinism

Three fundamental conceptions, that of a species, of adaptation and of evolution itself were fused in Darwin's theory of biological evolution. This chapter will introduce these ideas primarily to show the extent to which they had already coalesced at the time when Darwin began his work, and also to show how they converged on him before, during and immediately after the voyage of the *Beagle*. A subsidiary theme is the claim staked in all three ideas by the Christian, and particularly the Anglican, Church at the beginning of the nineteenth century.

The biological world suggests some sort of classification even to the most casual observer. There were three aspects of biological classification compounded in the concept of a species that was current in Darwin's youth. The first and most obvious aspect is the idea of an irreducible type, continuous within itself and discontinuous with other types. This is the normal vernacular 'species', the category 'cat' of which all individual cats are members, and which excludes all other creatures. The second aspect is the idea of a classificatory or 'systematic' hierarchy, in which all species can be ordered in relation to one another according to the degree of likeness. Similar species can be grouped into a genus, similar genera into a family and so on through a list of higher categories until very broad and general degrees of likeness such as 'plants' and 'animals' are being identified. The third and most elusive aspect is the idea of a hierarchy of endowment or some other value-laden notion. Species can be seen to exist in countless modes from elementary, motionless and insensible vegetables to subtle, active and sentient animals. At the bottom of this scale are organisms that can hardly be distinguished from the inorganic world, while the human species seems to occupy the pinnacle. All three aspects of the species concept, the notion of a type, and the position of each type in both a systematic and an evaluative hierarchy have been explicit in Western thought since Aristotle. Each made its own characteristic contribution to the received view of living things which dominated orthodox religious and biological thought in the pre-Darwinian nineteenth century.

The justifications for admitting the existence of the species as the fundamental unit of classification were many. Common observation insisted on it, and common experience added the further significant factor that individuals of a species mated successfully only with their own kind and generated only their own kind. The reality of such common objects of experience also earned a philosophical sanction from Platonic idealism. Individual cats impressed themselves on the mind not as individuals but as representatives of a kind: it was their essential catness, suggested only imperfectly by individual cats, which was the only logically operable subject for rational thought. Finally, the text of Genesis united common experience and Platonic philosophy; 'And out of the ground the Lord God formed every beast of the field and every fowl of the air; and brought them unto Adam to see what he would call them: and whatsoever Adam called every living creature, that was the name thereof.'

It is noticeable how closely entangled the concept of essential specific types is with the concept of permanence of species over time and through generations. The creation story in Genesis established an account for the origin of living things. Animals and plants were created in their severals kinds, and then persisted unchanged into the present by the power of reproduction. There were thus two formative principles for living things: the miraculous creative cause operating at the species level and the secondary cause of reproduction operating through individual members of a species. The core of Darwin's achievement in the *Origin of Species* was to challenge successfully this dualistic view of the origin of living things, and to replace it by the single definitely known formative principle, reproduction.

The classification of living organisms into empirically consistent groups by John Ray in the late seventeenth century, and by Linnaeus in the eighteenth century, had profound effects on pre-Darwinian biological and religious thought. The very fact that the assemblage of all species, no matter how many were discovered, was not chaotic demanded interpretation. If species had been individually created, what was the significance of this larger-scale organisation? In a conceptual world dominated by the creation myth the hierarchical classification of organisms received no explicit scriptural sanction. Systematic order in the organic world therefore became integrated into the post-Renaissance view of the natural world in general, as a complex machine whose operations were regular and governed by laws which established the proper relations between things. The order which prevailed in the organic world represented the workings of the divine legislator as much as did the laws of motion and gravity which governed the relations between physical objects. Various attempts were made in the early nineteenth century to abstract formal systems of living things from the empirical classifications, to find an order in classification itself that would be as rigorous and objective as the laws of physics.

Such 'schemes of nature' were always doomed by the variety of living things, hierarchically ordered, but not in any regular or symmetrical way. Some groups of organisms were incredibly various, others incomparably less so. The gulfs of difference that separated classificatory groups could be deep or shallow. Worst of all, even the gulfs that divided species from each other began to break down as animal and plant collections grew during the eighteenth century, and were subjected to progressively more searching systematic analysis. Linnaeus, who began his *Systema Naturae* (1735) with complete assurance that the species unit was absolute, ended his life uncertain about the validity of any of the normal sanctions of species status. Variation in anatomical structure clouded the issue of what was a species and what merely an insubstantial variety. Members of apparently different species could sometimes hybridise and produce fertile offspring. Linnaeus eventually settled for the created identity only of the next classificatory unit, the genus, in order to avoid the inconsistencies of a species-based scheme of nature. The fragmentation of the fundamental building blocks of organic creation by systematics in the eighteenth century formed a recurrent theme in the development of evolutionary biology.

One can see the two threads of systematic biology, the minute and refined analysis of wild species, and the attempts to enforce arbitrary order on empirical classifications, as opposing influences in the development of evolutionary biology. However, the search for the divine plan of creation which inspired the highest flights of folly in systematics also generated an enthusiasm for demonstrating God's earthly purpose by the investigation of the natural world. Scientific laws were to be seen as agents in the implementation of this purpose in the cosmos in general and on earth in particular. Natural phenomena were no longer miraculous in themselves: the relationships between things could be explained in terms of the laws of science. These were 'secondary' laws whose effects, if properly read, would illustrate the goodness of the divine legislator who had established them providentially to secure the comfort of man on earth. The parts of the whole natural system showed a mutual adaptation, guaranteed by the laws of science, that suggested a designing hand.

By the beginning of the eighteenth century Anglican orthodox opinion had virtually abandoned revelation in favour of the conformity of nature to man's needs as the primary source of evidence for the existence and attributes of God. As 'natural' theology developed, so every aspect of the fitness of things in the natural world, whether explicitly for the use of man or not, acquired evidential value. This theology was pervasive: in 1836, the year of Darwin's return from the *Beagle* voyage, the justly renowned Cambridge philosopher of science, William Whewell, was still able to consider 'the whole mass of the Earth from pole to pole, and from circumference to centre as employed in keeping a snowdrop in the position most suited to the promotion of its vegetable health'.

Such fatuities had already been challenged sixty years earlier by Hume's scepticism in the *Dialogues Concerning Natural Religion*: no claims about the excellence of present adaptation could refute the possibility of some antecedent condition that was not so good. Indeed, developments in geology might be argued to give evidence for the past existence of just such states. In the face of these pressures, Anglican natural theology retreated into the apparently impregnable fortress of biology. Only purposive design could fashion such complex mechanisms and such perfect adaptations to function as the human hand or the eagle's eye. The argument from design in the biological world was put with eloquence and cogency by William Paley, Archdeacon of Carlisle, in his *Natural Theology, or Evidence of the Existence and*

Attributes of the Deity collected from the Appearances of Nature (1802). When Darwin entered Cambridge in 1827, intending to become an Anglican priest, Paley represented the received opinion of the Church, and it was his works that Darwin could 'almost formerly have said by heart'.

There is no doubt that Darwin's deep immersion in natural theology at Cambridge did him a great service. The adaptation of plants and animals to their conditions of life was a real phenomenon which any evolutionary theory of biology would have to explain. In the hands of the natural theologians adaptation was more significant than the diversity of species: different animals and plants could be viewed as different ways of being adapted. If any biological adaptation could be explained without recourse to miracle the way was free to an explanation for all adaptation, and thence for the diversity of species.

Natural theology was above all a counsel of optimism, a belief in the essential goodness of the Creator. It was inevitable, therefore, that much of its effort, and its most elevating rhetoric, should have been devoted to an explanation for the paradoxical but insistent evidence for evil, pain and misery everywhere in the natural world, and nowhere more so than in God's paramount creation, mankind itself. The defence of God against the accusation of mismanagement of his affairs was based on the presumption that misery and pain indicated a higher and less evident purpose in the creative mind. If this argument was valid then the darker side of life was a necessary evil, and attempts to ameliorate man's suffering through social improvement were evidently contrary to divine decree and doomed to failure. In keeping with the spirit of natural theology, this position eventually found an empirical justification in the natural world. An Anglican cleric, Thomas Malthus, called attention to the inexorable war between creatures that followed from the universal tendency to multiply in number in a geometrical progression, and thus to exhaust the resources of the environment. In his *Essay on Population* (1798) Malthus argued that this principle applied equally to man and all other living things: it was a law of nature whose consequences could not be avoided. Man was therefore certain to suffer the indignities and miseries of over-population, and if consolation was to be sought it was only in the aspirations for a better life that such exigences must incite, not in any expectation that such aspirations could be universally fulfilled.

Malthus's *Essay* was a complex product of a complex period, but that it had its roots firmly based in natural theology and the concept of an ideal creation is perfectly clear. Nevertheless, the empirical generalisation that it contained, taken entirely out of its original context in 1838, supplied Darwin with the idea of the struggle for existence which forms one of the cornerstones of the theory of evolution.

The idea of evolution itself is, of course, the most profound and complex of the three central notions which together generated Darwin's theory of biological evolution. For Darwin, however, little of the history of the idea, or of the subtle and speculative forms it had taken before him, was of any importance. It came to him filtered by the critical eye of the geologist, Charles Lyell, when Darwin read the *Principles of Geology* on board the *Beagle*. For it was in geology that the idea of evolution left the domain of speculative philosophy where it had stayed for over two thousand years, and first entered into the domain of science.

The central idea in evolutionary philosophy is that the organisation of the world is in a state of flux. The idea becomes an object of scientific enquiry once it is assumed that changes in the organisation of matter are regular and rule-governed, that laws which describe the changing relations between things in known intervals of time apply with equal force in periods of time which are not directly experienced. By finding out what the causes of change are now, it should be possible in principle to explain how the world came to be the way it is. This is the principle of uniformity, introduced gradually as a dogmatic principle in geology during the eighteenth century. It is the foundation of all scientific theories of evolution, whether of the earth's surface, of living things, or of the cosmos.

It was inevitable that a geological science which looked at the surface of the earth as a mobile and changing structure and part of a mobile and changing cosmos should eventually come into direct conflict with theological limitations on the development of science. Historical geology, with its emphasis on slow and continued processes, introduced a new and almost limitless time-scale for the past evolution of the earth which recognised none of the miraculous and instantaneous events of the Mosaic creation story. Archbishop Ussher's chronology, calculated from the biblical text in the early seventeenth century, allowed 4004 years from the creation to the birth of Christ, while the rate of observable geological processes such as uplift, erosion and sedimentation obviously required many millions of years of earth history to explain the formation of great mountain ranges or deep river valleys. Furthermore the world picture which grew out of scientific evolutionary geology had no need for an imminent and interventionist deity: once the materials had been made available the evolution of the cosmos and the earth within it could reasonably be assumed to have looked after itself.

As the geologists expanded the time-scale of earth history, so they forced a re-evaluation of the concept of biological creation as well. The fossils of extinct organisms embedded in rocks that had suddenly become incredibly ancient could hardly even metaphorically be referred to Noah's flood. Evolutionary biology and evolutionary geology were thus inextricably linked, and eighteenth-century supporters of geological transformation were invariably committed to biological evolution as well. By the beginning of the nineteenth century evolutionary accounts of the origin of living things by descent from more primitive precursors were common currency, especially in France. Indeed by a happy, but probably not fateful, coincidence, Darwin's paternal grandfather, Erasmus Darwin, an ardent progressive in all matters, represented the concept of biological evolution to an English audience in his immense but popular *Zoonomia* (1794–6).

It is important to realise, however, that by the early nineteenth century there was a profound distinction between evolutionary ideas in geology and in biology. Evolutionary geology had by this time clearly moved out of speculative philosophy and into the scientific domain through its explicit recognition of the nature and effects of specific geological processes. Evolutionary biology, in contrast, was still essentially a speculative proposal, valuable as an explanatory device for certain purposes and philosophically more acceptable to progressive minds than the claustrophobic creationism of orthodox religion, but not supported by any understanding of the process or mechanism of evolutionary transformation. It is in this context that the idea of an evaluative classification came into its own, a progressive scale of nature leading from the lowest animalcules to the glory of mankind. Throughout the eighteenth century the scale of nature was arrayed along the dimension of time and identified with progress. The relationship between this metaphysical conception and the natural world remained hopelessly confused by the elusiveness of the evaluative principle, by the continuing existence of the primitive forms alongside the highest products of evolution, and by the entire absence of any mechanism for deriving one from the other.

As the methodological foundations of evolutionary geology became more secure, so its practitioners divested themselves of the encumbrance of an evolutionary biology whose scientific status was deeply suspect. Evolutionary geology directly contradicted scripture and only won its prolonged battle with theological orthodoxy by the irresistible strength of its arguments. The case for evolutionary biology was far weaker, and the professional geologists joined with the clerics in questioning its foundations.

It is appropriate to draw the complex threads of this chapter together through Lyell's attitude to the evolutionary theory of the French naturalist and geologist Jean Baptiste de Lamarck. Lamarck's theory of evolution was published in his book *Philosophie zoologique* in 1809, and clearly belongs in the tradition of speculative evolutionary biology. Its significance in the history of Darwin's theory is that it provoked a long and extremely able discussion of the case for biological evolution in the second volume of Lyell's *Principles of Geology*. Furthermore, Lamarck's theory of evolution fused all of the fundamental concepts to which this chapter has been devoted.

Lamarck, as naturalist, emphasised the extreme difficulty of distinguishing species from varieties and antedated Darwin by many years in denying that species had any objective reality. He also pointed out the delicacy and refinement of biological adaptation to circumstances, and noted that environmental changes could influence the structure and behaviour of plants and animals in a direction that was favourable to their existence.

As a geologist he emphasised the great geological and climatic changes to which the earth had been subjected over the geological time-scale. Finally, as a speculative evolutionist, he proposed a mechanism for the inheritance of structural change for which there was no evidence at all. Lamarck proposed that the evolution of animals was impelled by their recognition of new needs. This in turn provoked behaviour appropriate to the satisfaction of these needs, and behavioural change caused structural change which made the behaviour more efficient. Finally, the structural changes which developed during the lifetime of an animal in response to each mental initiative were inherited by the offspring. So the duck and the otter both acquired webbed feet by the enthusiasm of their ancestors for a watery existence, and by a similar process, impelled by different needs, man evolved from his ape-like ancestors.

Lyell recognised the strength of the biological and geological arguments, but was unable to accept the Lamarckian mechanism of evolutionary change as more than empty speculation. And so the greatest evolutionary geologist of the age dismissed all evolutionary biology until he was finally convinced that Darwin had indeed found a valid mechanism for the evolutionary process.

This, then, was Darwin's conceptual inheritance. The origin of species was the great unsolved question of his day, 'that mystery of mysteries, as it has been called by one of our greatest philosophers' (O 1). The philosopher was John Herschel, a physicist, whose work on scientific method Darwin had admired as an undergraduate. Paley's writings expounded the problem of biological adaptation to Darwin while he was still studying theology, to such effect that in the *Origin of Species* it was this above all that he saw as needing an explanation. Any theory of evolution 'would be unsatisfactory, until it could be shown how the innumerable species inhabiting this world have been modified, so as to acquire that perfection of structure and coadaptation which most justly excites our admiration' (O 3). Darwin's first exposure to the study of evolution came through the methodological purism of Lyell's *Principles*, read on the *Beagle* as he was beginning to unearth the fossils which first convinced him that organic evolution had indeed occurred.

Lyell's insistence that a scientific theory of evolution required a mechanism forced Darwin to consider this rather than any other aspect of evolution to be the crucial scientific issue. Less than two years after returning from the *Beagle* he had plundered Malthus for the missing link in his argument – and the mystery of mysteries was solved.

> I can entertain no doubt, after the most deliberate study and dispassionate judgement of which I am capable, that the view which most naturalists entertain, and which I formerly entertained – namely, that each species has been independently created – is erroneous. I am fully convinced that species are not immutable; but that those belonging to what are called the same genera are lineal descendants of some other and generally extinct species, in the same manner as the acknowledged varieties of any one species are the descendants of that species. (O 6)

The profound distinction that Darwin rightly perceived between his own evolutionism and that of his many predecessors lay in his solution to the crucial but subsidiary question of the mechanism of evolutionary change. The solution, christened natural selection by Darwin, was in fact an elementary syllogism whose premises were both simple and universally acceptable. The furious response which the *Origin of Species* received at the hands of its critics was surely a reaction to the logic of a simple argument from acceptable premises that led inexorably to an unacceptable conclusion.

3 Natural selection and the origin of species

Natural selection is the mechanism of evolutionary change. Unlike the processes of geological change it is not readily observable, but is inferred by argument from other kinds of observation. The argument is constructed from three apparently independent generalisations about the properties of organisms. These three generalisations then serve as premisses or axioms for a formal syllogism whose conclusion is a further generalisation about the properties of organisms. If the three axiomatic generalisations are valid, and if there is no other relevant valid generalisation that has not been considered; then the conclusion must also be true and should be observable.

The first generalisation is that individual members of any species vary somewhat one from another in manifold characteristics both structural and behavioural. Even the most rigid adherents of the immutability of species accepted this immediately ascertainable fact. 'No one supposes that all the individuals of the same species are cast in the very same mould' (O 45).

The second generalisation is that individual variation is to some degree hereditary, that is, transmitted from generation to generation. 'Perhaps the correct way of viewing the whole subject, would be, to look at the inheritance of every character whatever as the rule, and non-inheritance as the anomaly' (O 13).

The final generalisation is the Malthusian principle that organisms multiply at a rate which exceeds the capacity of the environment to carry them, with the inevitable consequence that many must die.

Every being, which during its natural lifetime produces several eggs or seeds, must suffer destruction during some period of its life, ... otherwise, on the principle of geometrical increase, its numbers would quickly become so inordinately great that no country could support the product. Hence, as more individuals are produced than can possibly survive, there must in every case be a struggle for existence, either one individual with another of the same species, or with the individuals of distinct species, or with the physical conditions of life. (O 63)

The principle of natural selection is a deductive consequence of heritable variation, multiplication and the struggle for survival.

If . . . organic beings vary at all in the several parts of their organisation . . .; if there be, owing to the high geometrical powers of increase of each species . . . a severe struggle for life; then, . . . I think it would be a most extraordinary fact if no variation ever had occurred useful to each being's own welfare . . . But if variations useful to any organic being do occur, assuredly individuals thus characterised will have the best chance of being preserved in the struggle for life; and from the strong principle of inheritance they will tend to produce offspring similarly characterised. This principle of preservation, I have called, for the sake of brevity, Natural Selection. (O 126–7)

How is natural selection a mechanism for evolutionary change? There are three points which may clarify this problem. The first is that natural selection is a process: each generation of organisms is subject to the selective impact of its environment and some of its members perish or fail to reproduce. The individuals which succeed in reproducing their kind will not be drawn at random from the population because selective pressures fall unequally on differently constituted individuals. If the environmental conditions for each generation of organisms are slightly different, as in the simple case of the slow development of a climatic change, such as an ice age, the individuals best capable of tolerating the change will tend to outbreed their less resistant cousins. Just as the geography of a valley is changed by the geological process of erosion by water, so the constitution of a population of organisms is changed by the persistent erosion of selection.

The second point is that natural selection and adaptation are obviously two sides of the same coin. An organism is said to be adapted to its conditions of life if it successfully passes the barrier between the generations. Although, to a biologist, the only general definition of 'adaptation' is the ability to reproduce, one can usually specify particular attributes which favour this ability, given the environment in which an organism lives. In the case of our hypothetical ice age, an attribute such as thick fur clearly promotes reproductive success, or fitness, under the prevailing cold conditions. But the concept of an adaptation is, in the light of natural selection, entirely conditional on the environmental pressures to which organisms are subjected. The thick fur which is so clearly an adaptation in the depth of an ice age, is equally clearly a liability as the ice age recedes. The 'adaptation' of the natural theologian was a static condition, while Darwinian natural selection redefines adaptation for each generation.

The third point is that natural selection is clearly understood to be a process that operates on a *population* of organisms. Individuals merely succeed or fail in reproduction: they are the cannon-fodder of the selective process. It is meaningless to say that individuals evolve: evolution is the change in the average constitution of a population of individuals as the generations succeed one another.

In later editions of the *Origin* Darwin introduced Herbert Spencer's phrase 'survival of the fittest' to encapsulate the idea of natural selection. This coinage has often provoked the accusation that nothing is really being asserted in the argument for natural selection: since fitness can only be defined by survival the phrase is a tautology. Only because the misunderstanding is so common is it worth trying to identify the problem a little more clearly. The easiest way to escape from the tautology is to remind oneself of the first axiomatic generalisation, that individual members of a species vary. Those that survive to reproduce are simply labelled after the event as 'fitter' than those that do not: natural selection is the *differential* loss of differently constituted individuals. It was not, perhaps, merely Darwin's adoption of Spencer's tautology that caused problems in understanding the point of the argument for natural selection. In his original formulation of natural selection (quoted above) Darwin used the phrase 'variations useful to any organic being', seeming to imply that the quality of usefulness was inherent in the variation, and that the environment was then 'exploited' by such well-endowed individuals. The question is simply, when does a variation earn its characterisation as 'useful' or 'harmful', when does an individual earn its characterisation as 'fit' or 'unfit'? The right answer must be, after selection, since the outcome of selection is the only criterion of usefulness or fitness. If Darwin did not make this point entirely clear, it was perhaps because he saw that the whole argument for natural selection seemed to involve a paradox, in that it is the *destruction* of individuals which is a necessary condition for adaptive or *constructive* change. If, however, he labelled the variations to be selected as 'useful' then the paradox seemed to go away. There really is no paradox, of course. Elimination of individuals in each generation *must* occur, whether they vary or not, because of the struggle for existence. In each generation the ability to reproduce is necessarily retained only by the survivors from the last, so the important thing in evolution is not to be selected against.

Darwin drew attention repeatedly to the complexity of the interrelationships between organisms and all aspects of their environment. In a striking simile which originated from one of the earliest notebooks he emphasised that the intensity of mutual competition between organisms was the dominant selective influence.

In looking at Nature, it is most necessary . . . never to forget that every single organic being around us may be said to be striving to the utmost to increase in numbers; that each lives by a struggle at some period of its life; that heavy destruction inevitably falls either on the young or old, during each generation or at recurrent intervals. Lighten any check, mitigate the destruction ever so little, and the number of the species will almost instantaneously increase to any amount. The face of Nature may be compared to a yielding surface, with ten thousand sharp wedges packed close together and driven inwards by incessant blows, sometimes one wedge being struck, and then another with greater force. (O 66–7)

This passage introduces a very important implication of

natural selection. Natural selection operates on the basis of individual reproductive performances: a 'useful' variation is one
which allows one individual to leave more offspring to the next
generation than another individual. Natural selection thus im-
plies 'selfish' organisms. The importance of this generalisation
to Darwin was enormous, since it discriminated between providential creation and evolution by natural selection.

> Natural selection cannot possibly produce any modification
> in any one species exclusively for the good of another species
> . . . If it could be proved that any part of the structure of any
> one species had been formed for the exclusive good of
> another species, it would annihilate my theory, for such could
> not have been produced through natural selection . . . I would
> almost as soon believe that the cat curls the end of its tail
> when preparing to spring, in order to warn the doomed
> mouse . . . Natural selection will never produce in a being
> anything injurious to itself, for natural selection acts solely by
> and for the good of each. (O 200–1)

We shall discuss briefly some of the important qualifications
to this generalisation in later chapters.

If mutual competition between organisms was a more important selective influence than mere climatic or geographical
change, then the direction of evolution of a species or of an
interacting group of species, became effectively impossible to
determine. The variability of each species will have consequences which reverberate far and wide through the whole interacting system. To ensure evolution it seemed unnecessary to
postulate any environmental variation except that caused by
other organisms:

> in several parts of the world insects determine the existence
> of cattle. Perhaps Paraguay offers the most curious instance
> of this; for here . . . cattle . . . have never run wild, though
> they swarm southward and northward in a feral [wild] state;
> . . this is caused by the greater number in Paraguay of a cer-
> tain fly, which lays its eggs in the navels of these animals
> when first born [and kills them]. The increase of these flies,
> numerous as they are, must be habitually checked by some

means, probably by birds. Hence, if certain insectivorous birds (whose numbers are probably regulated by hawks or beasts of prey) were to increase in Paraguay, the flies would decrease – then cattle . . . would become feral, and this would certainly greatly alter (as indeed I have observed in parts of South America) the vegetation: this again would largely affect the insects; and this . . . the insectivorous birds, and so on-wards in ever-increasing circles of complexity. We began this series by insectivorous birds, and we have ended with them. Not that in nature the relations can ever be as simple as this. Battle within battle must ever be recurring with varying suc-cess; and yet in the long-run the forces are so nicely balanced, that the face of nature remains uniform for long periods of time, though assuredly the merest trifle would often give the victory to one organic being over another. (O 72–3)

It is important to recognise this instance for what it is, namely a hypothetical scenario for fluctuations in frequency of different kinds of organisms that are all subtly related to each other, in-itiated by introducing into a dynamic equilibrium a single small change, the increase in insectivorous birds. It does not matter for the argument in principle whether the routes of influence suggested are in fact correct. The increase in navel flies may be primarily regulated by the availability of navels, and only secon-darily by insectivorous birds. The insectivorous birds may be primarily regulated by the availability of another species of in-sect or by disease, and not by predators. The key point is that to the extent that organisms are interactive, the effects of small changes in one organism are visited on all those other organisms with which it interacts, and so on ad infinitum.

Natural selection is an agency of adaptive change which oper-ates between generations. As the generations pass, so the living forms of a species come to differ progressively more and more from their ancestors. Although united by the bond of reproduc-tive continuity, it is easy to see how living organisms could eventually come to differ so much from their predecessors that the taxonomist would identify the living organism and its re-mote ancestor as different species. This, then, is an important part of the answer to the question, what is the origin of species?

But it is not, as Darwin rightly perceived, a sufficient answer unless all existing species represent the surviving members of separate lineages that have been kept separate since the origin of life on earth. It is not, therefore, an answer which excludes the intervention of an additional creative principle at some time past which called the primordial members of each separately evolving lineage into being; indeed it seems to require such a creative intervention.

The argument which Darwin developed to demonstrate that separate creative acts were unnecessary is the most complicated part of the *Origin of Species*, and, to the extent that it finally disposes of the creative principle, it must be regarded as the most crucial. What Darwin wished to demonstrate was that different living *species* were related by community of descent, that members of two such species, like two members of a single species were, however distantly, cousins by virtue of possessing a common ancestor. Darwin struggled for some time to find an argument which would prove that species would tend not merely to replace each other but also to branch in time, so that one past species should leave more than one descendant species to the present. In the *Origin*, Darwin introduced the concept of 'places in the polity of nature' to deal with the problem. We should nowadays refer to this concept as a 'niche'. Any species has a geographical 'range' or extension (e.g. from north to south). Some members will therefore be exposed to one set of conditions while other members are exposed to similar but distinct conditions. The complex interrelations between organisms which determine, for example, that an increase in insectivorous birds may by a series of intermediate steps alter the distribution of cattle, will guarantee that the checks to increase which operate on some members of a species are distinct from those that operate on other members. A species will therefore tend to adapt differently to the different conditions which prevail in different parts of its range. Groups of individuals of a single species which are adaptively differentiated to slightly different niches are called 'varieties'. To the extent that adaptive change in a variety influences the other organisms with which the variety interacts, by however complex a route, the process of adaptive change in the variety is accentuated, and its niche is further modified. As Darwin saw the matter, by the process of adaptive extension of a species into new niches, varieties should become progressively more clearly distinguished from the parental type and from each other until they would eventually acquire the characteristics of species themselves. There appeared to be no limit to the extent of divergence of character which could be achieved by means of sequential adaptation of the different sections of a species to different niches.

The 'polity of nature' presents an infinity of places into each of which a variety can attempt to 'wedge' itself by divergence from its ancestors. Darwin made it clear that he also saw this process resulting in a persistent tendency to fill all the spaces. Because of time, variation and reproductive potential, the polity of nature must always tend to maximise its occupancy, and, since each type of organism essentially creates a niche for another type, heterogeneity is a deducible consequence. The existence of vegetable life 'creates' a niche for the eaters of vegetables; the eaters of vegetables 'create' a niche for their own carnivorous predators. A crude example, but adequate to illustrate the principle, for 'more living beings can be supported on the same area the more they diverge in structure, habits, and constitution'. Maximum occupancy is thus ensured by maximum heterogeneity, so the divergence of species seems also to entail the multiplication of species. However, to the extent that adaptive evolution of one type bears down on the conditions of life of another, not just individuals but whole species are subject to extinction by competition. Species may multiply but they will tend to do so at the expense of other species. Extinction may be avoided by adaptive change, but it may not. No rule guarantees the indefinite persistence of any organism or group of organisms.

In the form in which Darwin expresses his argument, there can be no doubt that adaptive divergence of character within a species in different parts of its range is a legitimate consequence of the fundamental premises on which the entire theory of evolution is based. Furthermore, if adaptive divergence be allowed to any small extent there is no logical limit to its indefinite continuation. There is, however, a very fundamental gap in the argument which Darwin himself certainly saw but never adequately resolved, namely that it really accounts only for divergence *within* a species, and still leaves open the crucial issue of how one species can evolve into more than one species. The central dilemma concerns the definition of a species. We have seen that the vernacular definition of a species unequivocally includes the notion of discontinuousness from other species. The cat species does not seem to be truly continuous with other similar species which might reasonably be argued to be, on the principle of divergence, its 'cousins'. Yet there is in Darwin's argument no agent or principle which seems capable of introducing a true discontinuity between two diverging lines of a continuous species. One section of a species may be heritably equipped to reproduce in one extreme part of its range, and another section differently equipped to deal with another extreme part of the range. But the argument for adaptive change applies equally to all individuals which intervene between the two extremes and, if the range is continuous, so the extreme forms of a species must merge into each other through an indefinite chain of less 'extreme' variants. There is, indeed, a powerful influence which acts in opposition to the divergent tendency, and that is the influence of sexual reproduction. By combining the heritable characters of two individuals the sexual process produces progeny with characters, on average, intermediate between their parents. Thus the 'normalising' influence of sexual reproduction must reduce the rate at which a new niche can be occupied by a potentially divergent section of a larger population with which it is in sexual continuity. A species is indeed, as in the vernacular definition, continuous within itself and, despite the divergent tendency correctly identified by Darwin, inherently conservative.

The necessary additional postulate which permits a species truly to fragment into two species is reproductive isolation between potentially divergent sections. So far, it has proved extremely difficult to construct a deductive argument from adaptive divergence alone which can create reproductive isolation out of a pre-existing state of reproductive continuity. Reproductive isolation seems to be something which a variety of a species cannot achieve for itself. At present it is widely, if not universally, held that a physical barrier has to be introduced between two varieties before they can become two species. Darwin did not by any means ignore the importance of isolation in the process of speciation, but he seems gradually to have lost sight of its crucial place in the argument. In his earliest notebooks on transmutation of species, isolation was repeatedly invoked as an important factor in speciation, and in these early notes Darwin also realised the problems that follow the absence of isolation: 'Those species which have long remained are those which have wide range and therefore cross and keep similar. But this is difficulty: this immutability of some species.'

In the *Origin*, Darwin again drew attention to the opposed influences of the tendency towards divergent adaptation and the sexual continuity of a species:

But if the area [occupied by a species] be large, its several districts will almost certainly present different conditions of life; and then if natural selection be modifying and improving a species in the several districts, there will be intercrossing with the other individuals of the same species on the confines of each. And in this case the effects of intercrossing can hardly be counterbalanced by natural selection always tending to modify all the individuals in each district in exactly the same manner to the conditions of each; for in a continuous area, the conditions will generally graduate away insensibly from one district to another. (O 102–3)

However, even in the last edition of the *Origin* (1876), Darwin stuck to his opinion that while isolation may indeed promote speciation, the process could nevertheless occur without it. This insistence is the more surprising in view of the fact that not only had Darwin initially viewed isolation as a virtual requirement for specific divergence, but the same point had been taken up directly, and in the light of the theory of evolution, by the German naturalist Moritz Wagner (1813–87).

The situation is rather odd. In the absence of isolation Darwin's theory of evolution is truly inadequate as a mechanism of speciation. Furthermore, he understood isolation, had observed it, had noted its favourable consequences, and noted the unfavourable consequences of its absence. Finally, the issue had been brought explicitly to his attention by a distinguished naturalist strongly predisposed to evolutionary views and at a time when Darwin was preparing the most extensive of the his many revisions of the *Origin*. It may be that the contingent aspect of isolation was what offended Darwin. Variation, heredity and multiplication were obviously intrinsic to living things: they could be used virtually as formulae to derive evolution. Isolation was merely something that *might* happen to a section of a population, by its opportunistic arrival on an island, by the intrusion of a mountain barrier or by some other messy and illogical process. Since such messy and illogical processes certainly *did* happen, Darwin had no quarrel with the idea of isolation, indeed it served him well in explaining some of the observable facts of geographical distribution. But he never quite accepted that the formation of new species was not also intrinsic to living things as a necessary consequence of natural selection alone.

In this chapter we have dealt with the core of Darwin's theory of evolution. From the generally agreeable premisses that organisms tend to vary, however slightly, in all aspects; that variations, however slight, tend to be inherited; and that more organisms are born than can survive to reproduce, we can deduce adaptive change in a species without necessary direction and without limit in extent. The organisms alive today are all the modified descendants of organisms alive in the past. Furthermore, given populations extended in space and given the opportunity for reproductive isolation, we can infer additionally that the discontinuous clusters of similar organisms called species are in turn the modified descendants of a smaller number of previously existing species. Since adaptive modification has no necessary limit, there is no limit to the amount of difference between two organisms which can be explained in terms of the inheritance of accumulated variation. Hence, there appears to be no *a priori* argument which excludes the common ancestry of all living things; that is, things which share the properties of variation, heredity and multiplication on which the argument rests.

Analogy would lead me one step further, namely, to the belief that all animals and plants have descended from some one prototype. But analogy may be a deceitful guide. Nevertheless all living things have much in common . . . Therefore I should infer from analogy that probably all the organic beings which have ever lived on this earth have descended from some one primordial form, into which life was first breathed.

(O 484)

4 The evidence for evolution by natural selection

The *Origin* is, as Darwin said, one long argument, but it extends well beyond the theoretical core described in the last chapter. If the core of the argument is correct, then certain consequences should be observable in the real world. If such implications of the argument are indeed observed, they may in turn be used as evidence in its favour. Much of the *Origin* is devoted to a virtuoso organisation of facts, most of them widely admitted by Darwin's contemporaries, presented as implications of the essential theoretical framework. The scope of the theory of evolution is such that most of the 'facts' which Darwin used were really themselves large generalisations. Compared with his later works, the *Origin* is tightly argued, and I disagree with T. H. Huxley's view of it as 'a mass of facts crushed and pounded into shape, rather than held together by the ordinary medium of an obvious logical bond'.

Natural selection is presented as the mechanism of evolutionary change. Darwin's first task was to show that in domesticated animals and plants a process formally identical to natural selection had produced a phenomenon formally identical to evolution. That process was the selection by animal and plant breeders of variants, however slight, that caught their fancy. 'Artificial' selection could produce true-breeding varieties that suited whatever purpose the breeder had in mind. In other words, selection was related to adaptation.

One of the most remarkable features in our domesticated races is that we see in them adaptation, not indeed to the animal's or plant's own good, but to man's use or fancy . . . We cannot suppose that all the breeds were suddenly produced as perfect and as useful as we now see them; indeed, in several cases, we know that this has not been their history. The key is man's power of accumulative selection: nature gives successive variations; man adds them up in certain directions useful to him. (O 29–30)

Furthermore, the scale of the changes that could be produced by this process among domesticated varieties of a single species (of which the pigeon was Darwin's favourite example) far exceeded the normally accepted definition of a species as a group or organisms that look essentially similar.

> Altogether at least a score of pigeons might be chosen, which if shown to an ornithologist, and he were told that they were wild birds, would certainly, I think, be ranked by him as well-defined species. Moreover, I do not believe that any ornithologist would place the English carrier, the short-faced tumbler, the runt, the barb, the pouter, and fantail in the same genus; more especially as in each of these breeds several truly-inherited sub-breeds, or species as he might have called them could be shown him . . . May not those naturalists who . . . admit that many of our domestic races have descended from the same parents . . . learn a lesson of caution, when they deride the idea of species in a state of nature being lineal descendants of other species? (O 22, 29)

The intervention of the conscious decision of the breeder might seem to diminish the force of the example, so Darwin pointed out that a change in character could occur by selective breeding even when there was no specific intention of altering the breed. Furthermore, the criteria for selection might change as the breed changed or threw up some unusual variant.

> No man would ever try to make a fantail, till he saw a pigeon with a tail developed in some slight degree in an unusual manner . . . But to use such an expression as trying to make a fantail, is, I have no doubt, in most cases utterly incorrect. The man who first selected a pigeon with a slightly larger tail, never dreamed what the descendants of that pigeon would become . . . (O 39)

This passage illustrates an additional point which Darwin used later to confront a common objection to his evolutionary views, that structurally perfect and highly adapted structures can only work when they are complete. How, then, could they be selected through the postulated intermediate stages? In the complex process of artificial selection lay one form of answer, namely that selection may seem goal-directed when viewed with hindsight, while in fact the standards required for selection may have been different at intermediate points.

The phenomena associated with domestication constitute a virtually complete 'experimental' verification of the principle of natural selection. Had 'improvement' never been attempted until the publication of the *Origin*, man's ability to mould the character of living things intentionally by the application of selection to inheritable variation would have been seen as a stunning vindication of the principle. In fact, the consequences of artificial selection had already been assimilated into a different view which took the fixity of species for granted. This view was that artificial selection could proceed to the varietal stage, but, by immutable law, not beyond. Varieties or races were no less members of their parental species after than before selection.

In one critical respect, therefore, the 'experiment' of artificial selection was incomplete. Although the evolution of form well beyond any normal conception of a species was clearly established, the 'experiment' had not yet demonstrated that artificial selection could generate the functional difference between species that had for a long time been offered as an additional and crucial criterion, namely mutual infertility. Darwin devoted a whole chapter of the *Origin* to the problems of hybridisation and infertility. His approach was to point out, with numerous examples, that the outcome of experimentally attempted hybrid unions between species was unpredictable.

> Now do these complex and singular rules indicate that species have been endowed with sterility simply to prevent their becoming confounded in nature? I think not, for why should the sterility be so extremely different in degree, when various species are crossed, all of which we must suppose it would be equally important to keep from blending together? Why should the degree of sterility be innately variable in the individuals of the same species? Why should some species cross with facility, and yet produce very sterile hybrids; and other species cross with extreme difficulty, and yet produce fairly fertile hybrids? Why should there often be so great a differ-

ence in the result of a reciprocal cross [i.e. male A with female B and vice versa] between the same two species? Why, it may even be asked, has the production of hybrids been permitted? to grant to the species special power of producing hybrids, and then to stop their further propagation by different degrees of sterility, not strictly related to the facility of the first union between their parents, seems to be a strange arrangement. (O 260)

Thus the providential guarantee that species should maintain their created identity dissolved into a conflicting jumble of observations which Darwin was able to make little enough sense of. His considered view was that the degree of mutual infertility between two organisms was essentially accidental, a fortuitous consequence of general physiological or behavioural divergence between species as they became adapted to different niches over a long period of time. Darwin rightly identified the ability to reproduce successfully as a fragile quality that could easily be upset by physiological changes in the reproductive system. As species diverged, so the whole mechanism of the reproductive system could be subtly altered as an indirect consequence of evolutionary changes that were affecting other parts of the organism. Obviously the reproductive systems of males and females within a species would be kept fully compatible by selection, but there was no discernible reason why selection should maintain compatibility between males and females of two different species. As they evolved independently of each other, so their reproductive systems would gradually drift away from mutual compatibility. Why then were domesticated varieties of a species, however much modified in appearance, always able to breed together? Darwin had difficulty with this problem, which was brought up repeatedly by his friends as well as his critics after the publication of the *Origin*. His answer to it was that the selection exercised by man in the formation of domestic breeds was applied only to 'external' characters, and that the breeder could not produce 'recondite and functional differences in the reproductive system' of the type proposed to cause infertility between wild species. Artificial selection was rapid and superficial, while natural selection was slow and deep.

This argument is partly correct but misses the key point. Darwin's problem throughout the discussion of hybrid infertility was that he was unable to see how mutual infertility itself could be a character evolved by natural selection, since 'it could not have been of any direct advantage to an individual to breed poorly with another individual of a different variety, and thus to leave few offspring.' He therefore had to leave mutual infertility as an accidental physiological consequence of other changes. The solution to this problem is again to be found in the phenomenon of reproductive isolation. As discussed in chapter 3, adaptive divergence of two varieties of a species is limited by the 'averaging' effect of sexual reproduction. If a geographical barrier physically prevents sexual intercourse between two varieties, then adaptive divergence can go much further and much faster. If, however, the geographical barrier is transient, and the two rapidly diverging varieties are thrown together again, the persistent ability to interbreed may actually become a disadvantage, since mongrel progeny will not be well adapted, on average, to the niches of either parent. In such a case the selective advantage lies with the individuals that breed only with their own variety. As a result, 'natural aversion', as Darwin called the behavioural disinclination to hybridise in animals, would be evolved as surely as any other adaptive character. An analogous physiological separation would evolve in the reproductive systems of plant varieties.

It is odd that Darwin did not develop this argument, but it is consistent with his failure to identify reproductive isolation as a necessary condition for the formation of species out of varieties. Darwin's demolition of divinely endowed sterility between species was thus a double-edged sword. He was able to use the evidence that species could frequently, if inconsistently, hybridise to dispose of the conventional 'functional' criterion of a species. But by failing to see the importance of reproductive isolation in speciation he was left with a weak and unconvincing explanation for the fact that different species did not, in general, hybridise, and concluded unnecessarily that inability or disinclination to mate successfully with a distant relative must be exempt from the otherwise pervasive action of natural selection.

The facts of domestication and hybridisation enabled Darwin

to discard either morphological or functional criteria for the definition of species. He was then free to adopt the most extreme opposite view, essentially identical to Lamarck's view of fifty years earlier:

> it will be seen that I look at the term species, as one arbitrarily given for the sake of convenience to a set of individuals closely resembling each other, and that it does not essentially differ from the term variety, which is given to less distinct and more fluctuating forms. The term variety, again, in comparison with mere individual differences, is also applied arbitrarily, and for mere convenience sake. (O 52)

A species became a group of evolving individuals distinct only for the incidental reason that it was seen in an extremely small time frame. Its true context was its past history and its future destiny as a part of the total connected structure of living organisms. For Darwin, such an extreme redefinition of the species was of immense polemical value in debate with a scientific community largely committed to species as external and essential realities. The disadvantage that we see clearly today is that it utterly ignores the real reproductive discontinuity between species in the wild, which is the fundamental source of isolation and divergent evolution.

Domestication and hybrid fertility provided evidence for the theory of evolution and against the fixity of species, by experimental intervention in the natural world. But it was Darwin's interpretation of the directly observable facts of wild nature which above all carried the day for evolution. In each of four major areas of investigation, geological, geographical, taxonomic and embryological, information was already available to attract the defensive attention of adherents of miraculous creation. Much of the argument Darwin used for evolution by natural selection was thus presented dialectically as he compared the explanatory power of his own views with the 'explanations' of the creationists.

In geology, the theory of evolution could explain two agreed facts that had given the scriptural view of creation a hard time. First, evolution must normally be slow. The existence of highly evolved and complex creatures with long generation times could guarantee this conclusion, since a generation was the shortest interval in which selection could act. An elephant could not conceivably be evolved in a hurry with a generation time of thirty to sixty years. Descent with modification clearly implied that the world must be old, and as geology had definitively proclaimed, it was indeed incredibly old. Secondly, descent with modification equally plainly required that fossils should, in general, be different from living organisms, and so the fossil record declared.

Darwin did not waste much space on these elementary points; their evidential value for an evolutionary biology had been understood for generations. His main concern was to counter the objections that had already been raised to the obvious interpretation of the evidence. The prevailing compromise between geological evidence of ancient and peculiar fossils and the biblical story of creation was based on the fact that fossil-bearing sedimentary rocks were arranged in superimposed strata with obvious discontinuities between them. Fossils in each stratum were largely characteristic of that stratum, and, particularly in the older rocks, the fossil discontinuities were very great indeed. Whole great groups of organisms (ammonites or dinosaurs) appeared or disappeared virtually at once. Most telling of all was that very advanced creatures such as fishes could already be found in the very oldest strata. To the creationists these discontinuities told of catastrophic disturbances of the natural order resembling Noah's flood. Whole systems of creation had been swept away and replaced by new ones. Where were the intermediate forms of life?

Uniformitarian geology did not deal in catastrophes, and Darwin did not deal in creation. His approach was to attack the proposition that the geological history of the earth was complete, and to infer that a discontinuity between fossil-bearing strata was simply a gap in the record. Only a small part of the surface of the earth had been explored geologically, and virtually none of it with any completeness, since entirely new fossils were being discovered every year. Furthermore, the preservation of a fossil at all depended on the concurrence of a number of favourable circumstances. A fossil-bearing sediment could be deposited only during periods of slow land subsidence: at other times sedimented material would be removed again by erosion. Subsidence was a local event in global terms. No sedimentary bed could therefore be considered complete unless its full history was known, but geological and climatic change conspired to make this unknowable. Since organisms show constant adaptation to their environment, geological discontinuities, implying environmental change over unknowably long time periods, should necessarily be accompanied by discontinuities in the fossil record.

The catastrophist argument depended on simultaneity. The whole global slate was wiped clean and a new organic creation superseded the older creation. The argument was most weighty when very large discontinuities were seen. The sudden appearance of abundant novel forms of life at the beginning of one great geological division and the sudden extinction of previously abundant forms at the end of another division were well known. Darwin dealt with the last resort of creationism in geology in three ways. First, he reiterated the general point about geological discontinuities, that they implied time and change. Secondly, although organic discontinuities were often very large, they were usually not complete. As geological research proceeded, so it tended to blur the sharp edges. Finally, he points to a purely 'Darwinian' reason why major floral or faunal groups should seem to appear suddenly:

> It might require a long succession of ages to adapt an organism to some new and peculiar line of life, for instance to fly through the air, but that when this had been effected, and a few species had thus acquired a great advantage over other organisms, a comparatively short time would be necessary to produce many divergent forms, which would be able to spread rapidly and widely throughout the world. (O 303)

Darwin was now free to read such of the record as existed as if it were the book of evolution. Not only was there clear evidence for evolution to be found, but much of it was most easily explicable in terms of evolution by natural selection. First, the rate of evolutionary change clearly varied: some organisms, always marine, remained virtually unchanged from the earliest known fossil-bearing rocks till the present, while others, especially terrestrial, changed rapidly. Yet some change was an invariable rule. Why should evolutionary rates be variable? The argument for rapid change following the evolution of a novel capacity, such as flight, has already been given. But now Darwin puts the whole case more generally.

> I believe in no fixed law of development... The process of modification must be exceedingly slow. The variability of each species is quite independent of that of all others. Whether such variability be taken advantage of by natural selection, and whether the variations be accumulated to a greater or lesser amount... depends on many complex contingencies, -on the variability being of a beneficial nature, on the power of intercrossing, on the rate of breeding, on the slowly changing physical conditions of the country, and more especially on the nature of the other inhabitants with which the varying species comes into competition. Hence it is by no means surprising that one species should retain the same identical form much longer than others; or, if changing, that it should change less. (O 314)

Extinction was one of the most obvious features of the fossil record, and Darwin rightly saw it as the most compelling evidence for the action not just of evolution but of natural selection.

> The theory of natural selection is grounded on the belief that each new variety, and ultimately each new species, is produced and maintained by having some advantage over those with which it comes into competition; and the consequent extinction of the less-favoured forms almost inevitably follows... Thus the appearance of new forms and the disappearance of old forms... are bound together... But whether it be species belonging to the same or to a distinct class, which yield their places to other species which have been modified and improved, a few of the sufferers may often long be preserved, from being fitted to some peculiar line of life, or from inhabiting some distant and isolated station, where they have escaped severe competition... We need not

marvel at extinction; if we must marvel, let it be at our presumption in imagining for a moment that we understand the many complex contingencies, on which the existence of each species depends. (O 320–2)

Extinct organisms could either represent 'linking fossils', the extinct precursors of groups of organisms alive today, or they could be extinguished forms of life which left no descendants to the present. On the whole it was the class of 'linking' fossils which attracted most of Darwin's attention, since their existence was specifically implied in the theory of evolution, and not accounted for by any subsidiary argument from the idea of special creation. Even in the mid-nineteenth century, the known fosssils immediately suggested the presence of linking forms. It was, indeed, a guiding principle of geological investigation that the more nearly adjacent two strata were in a sedimentary series, the more similar were the fossils they contained. The principle was equally applicable whether the fossils in question were relatively recent or very ancient, and whether two fossil beds were compared or any one fossil bed with the organisms alive today. The general point was that the guiding taxonomic principle of degree of likeness between organisms could be extended 'vertically' in time just as it extended 'horizontally' among living organisms. Extinct and living species

all fall into one grand natural system; and this fact is at once explained on the principle of descent. The more ancient any form is, the more, as a general rule, it differs from living forms. But, as Buckland long ago remarked, all fossils can be classed either in still existing groups, or between them. That the extinct forms of life help to fill up the wide intervals between existing genera, families, and orders, cannot be disputed. For if we confine our attention either to the living or to the extinct alone, the series is far less perfect than if we combine both into one general system. (O 329)

As the fossil record grew during the nineteenth century, so the greater or lesser divisions between forms separated in time, or living now but separate in structure, tended to 'dissolve'. But the imperfections in the record enormously diminished the probability that any particular form unequivocally linking two subsequent forms should ever be discovered.

One of the most spectacular 'missing link' fossils of all was still undiscovered in 1859. In 1861, the fossil *Archaeopteryx* was discovered in an ancient rock formation in Germany and was immediately recognised to be as truly intermediate between the extinct Dinosaurs and the modern true birds as anybody could have wished. Darwin duly included a reference to *Archaeopteryx* in the later editions of the *Origin*, but the general point was already so clear that it did not stand or fall on the discovery of one new fossil.

A special case of close similarity between the organisms in closely consecutive strata was that of the recent fossils found in a particular area and the living species found there. This special case also has a special place in the history of Darwin's theory of evolution.

> When on board H.M.S. 'Beagle,' as naturalist, I was much struck with certain facts in the distribution of the inhabitants of South America, and in the geological relations of the present to the past inhabitants of that continent. (O 1)

The sloths and armadillos are rather odd mammals characteristic of South America. Darwin discovered fossil bones in relatively recent sedimentary rocks belonging to gigantic extinct animals that plainly belonged to these groups. Why should any miraculous creative power have tried sloths and armadillos once in South America, and then again created closely related types in the same region and nowhere else? Why should such a creative power have acted similarly in Australia where the bones of extinct mammals found in caves are all closely related to the equally peculiar living Australian mammals? On the theory of descent with modification such correlations were expected. A recent ancestor will tend to die geographically closer to a descendant than will a more distant ancestor, since the positions occupied geographically by a species will depend on the prevailing conditions. As the prevailing conditions change through geological time, so the range of the species will change.

From the perception of motion through time which he acquired from Lyellian geology, Darwin inferred a concomitant motion in space. The more recent a common ancestor, the closer its divergent products should be to each other in space. The theory of evolution therefore predicted that species belonging to a particular group should be more alike the closer they were geographically, regardless of environmental conditions. On any theory of providential creation, organisms should be more like each other the more similar the environment and regardless of geography.

Darwin's treatment of this potential source of evidence for the theory of evolution concentrated first on the general conformity of observation to expectation over the great continental land masses.

> In considering the distribution of organic beings over the face of the globe, the first great fact which strikes us is, that neither the similarity nor the dissimilarity of the inhabitants of various regions can be accounted for by their climatal and other physical conditions . . . There is hardly a climate or condition in the Old World which is not paralleled in the New . . . Notwithstanding this parallelism . . ., how widely different are their living productions! (O 346–7)

Secondly, proximity seemed to be defined not by mere horizontal distance but by the presence or absence of barriers across which evolving and diverging groups would find difficulty in passing. In other words, the greater the isolation inferred on geographical grounds, the greater the average dissimilarity between organisms. Obviously the separation of the great continental land masses was a case in point, where the isolation was achieved by large tracts of water. But on a smaller scale, lesser barriers also seemed to be associated with relative dissimilarity, even in a single continent,

> for on the opposite sides of lofty and continuous mountain-ranges, and of great deserts, and sometimes even of large rivers, we find different productions . . . Turning to the sea, we find the same law. No two marine faunas are more distinct, with hardly a fish, shell, or crab in common, than those of the eastern and western shores of South and Central America;

yet these great faunas are separated only by the narrow, but impassable, isthmus of Panama. (O 347–8)

Finally, each major geographical unit did not merely possess characteristic species, but also, within any taxonomic group, the species of each geographical unit were more closely related to each other than they were to members of the same group in another geographical unit. There are monkeys in both Old and New Worlds, but New World monkeys are more similar to each other than they are to any Old World monkey.

We see in these facts some deep organic bond, prevailing throughout space and time, over the same areas of land and water, and independent of their physical conditions ... This bond, on my theory, is simply inheritance, that cause which alone, as far as we positively know, produces organisms quite alike, or, as we see in the case of varieties nearly like each other. (O 350)

If, however, relative likeness implied common ancestry, then the presence of even somewhat similar organisms on the opposite sides of great barriers implied that individuals must at some time have been able to cross these barriers.

Darwin viewed the great continental land masses as having existed more or less permanently in their present forms. In this century we have come to realise that in very distant geological epochs the great continents were united, and have since separated by 'continental drift' into their present positions. Continental drift does, in fact, explain some aspects of the geographical distribution of certain ancient groups of organisms, but Darwin was more concerned with more recent periods of evolutionary time. He was therefore forced to demonstrate the feasibility of other means of dispersal by which organisms could pass between isolated regions.

He determined experimentally that seeds of land plants were able to survive prolonged immersion in salt water, from which he could argue how far a viable seed might be transported by ocean currents. He invoked earth attached to floating tree stumps washed down rivers, the mud attached to birds' feet, the seeds carried in the crops of birds, the immense distances

travelled by icebergs carrying fragments of earth and other terrestrial matter. He showed how terrestrial animals such as snails
can survive immersion even in salt water by forming a hard
membranous diaphragm over the mouth of the shell. The dispersal of freshwater animals posed a special problem, but here
too Darwin's experiments demonstrated the principle that he
needed. He showed that newly hatched freshwater snails would
attach themselves firmly to ducks' feet and live for 'from twelve
to twenty hours; and in this length of time a duck or heron
might fly at least six or seven hundred miles, and would be sure
to alight on a pool or rivulet if blown across sea to an oceanic
island or to any other distant point' (O 385).

Dispersal, isolation and selection were nowhere more perfectly exemplified than on oceanic islands. The biology of the inhabitants of the Galapagos Archipelago was so extraordinary
and so suggestive that in describing it in the *Journal of Researches* Darwin was scarcely able to conceal his evolutionary
views. In the Galapagos, 'both in space and time, we seem to
be brought somewhat near that great fact – the first appearance of
new beings on this Earth' (J 378). The point was this. Geologically the Galapagos were clearly of relatively recent volcanic origin, yet they contained numerous species of plants and animals
that were unique to the islands: indeed there were even species
that were unique to individual islands. Darwin did not realise
the full extent of this island speciation because he mixed
together all the specimens of a group of birds, now known as
'Darwin's Finches', that he had collected from different islands. But it was not just that the Galapagos housed many
unique species: the unique species were quite obviously closely
related to well-known species that Darwin had already seen on the
South American sub-continent, the closest land-mass, 600 miles
to the east. Finally, the representation of different kinds of
animals and plants was peculiar. There were, for example, no
mammals and no trees. Yet the places normally occupied by
these missing groups were, however incompletely, filled by
grotesque variations of animals and plants whose closest mainland relatives led quite different lives. Analogous peculiarities
marked other oceanic islands such as Hawaii where ferns grew
into trees, and Mauritius and New Zealand, where huge flight-

less birds grazed on the mammal-less ground.

The bizarre features of island life could in no way be 'explained' on the basis of a rational process of creation, while the very isolation of the islands seemed to require that here, if anywhere, creation must have acted. The theory of evolution, on the other hand, asserted that the inhabitants of oceanic islands must originally have come from somewhere else. For Darwin, oceanic islands presented the results of what seemed to be an exemplary natural experiment in evolution. All the anomalies could be explained if the original population had been drawn virtually at random from the neighbouring mainland, selected only by the accidents of occasional dispersal, allowed to colonise an untenanted tract of land and given a period of geological time in which to evolve. The great merit of oceanic islands for Darwin's argument was the simplicity of the relationship between organisms and their environment. In a fully occupied land the evolutionary fate of a group of organisms is essentially impossible to predict because of the complexity of biological interactions. On oceanic islands competitive pressure is incomparably less complex because of the paucity of inhabitants and the abundance of unfilled niches. Each oceanic island thus presents a slightly different solution to common problems of adaptation to a relatively simple and reproducible environment. To a certain extent, therefore, evolution on oceanic islands is determinate, its duration known, and its direction predictable.

The distribution of animals and plants in time and space was certainly the most compelling evidence Darwin used in support of the generality of the theory of evolution. Great clusters of 'brute facts' could be organised into a single satisfying conceptual framework that no creationist view could even attempt. The same fluent explanatory power of the theory of evolution was as obvious in Darwin's treatment of the problem of classification. Why should all organic beings be found 'to resemble each other in descending degrees, so that they can be classed in groups under groups'? (O 411).

The ancient idea of a Natural System of classification implied that if the 'right' points of likeness and difference could be identified, all animals and plants could be allocated to their proper place. The classification was, on this view, non-arbitrary and pre-imposed. Since animals and plants were extremely various, and presented many points of similarity or difference whose emphasis could lead to paradoxical arrangements, the search for the Natural System led to the attempt to find points of 'classificatory value', single diagnostic characters which would permit an unambiguous tree of defined choices leading finally to the individual species. This procedure was normally successful and still forms the pragmatic basis for classification. Darwin's assault on the traditional theory was directed against the idea that characters of high classificatory value could be established as if they possessed a real existence independent of the units to be classified. If organisms were related to each other by descent, a character of classificatory value was simply one whose modifications under selection were not so extreme that the similarity due to inheritance was lost.

Since, despite their aspirations, classifications were normally pragmatic, the levels of the hierarchy of classification, species, genus, family, order and class, were themselves also only levels of convenience and did not reflect any necessary condition of the living world. The taxonomic level was defined simply by the most convenient arrangement for the existing state of knowledge.

Thus, despite the obvious taxonomic order in the natural world, all attempts to define it by a scheme more rigorous and objective than the organisms to be defined seemed doomed to failure. The Natural System of the creator dissolved into a descriptive exercise whose only analytic content consisted of defining groups *post hoc*. Yet groups there evidently were, and hierarchically organised. The theory of evolution directly predicted a hierarchical arrangement of taxonomic groups as a consequence of sequential divergence from common ancestral forms, with extinction. It predicted the kind of arrangement, and additionally predicted the difficulties, that attended any attempt to establish non-arbitrary definitions of any kind. Dominant groups of organisms had repeatedly fragmented into large numbers of subordinate groups. Extinction of early and intermediate forms ensured that all groups would be divided by gulfs of difference whose depth would be some function of the time interval between the present and the common ancestor, and of the rate of divergence. And since species continuously emerge from divergence between varieties, attempts at an absolute definition of the species category itself were doomed to failure.

In the search for characters of absolute taxonomic significance, certain rules had been found to be of value. Relations between parts, i.e. anatomical organisation, were in general more informative than structure seen in relation to function: two organisms adapted to an approximately similar way of life must necessarily present similar characters. Birds, bats and insects all fly and all have wings, but as a taxonomic character it was equally apparent that such 'analogical' resemblances were useless. From a deeper organisational point of view, however, the forelimbs of birds, reptiles and mammals showed anatomical resemblances, so that it was possible, with greater or lesser qualification, to give the same names to the bones composing each of these limbs despite great functional differences. Such structural correlation, known as 'homology', was seen most emphatically in embryonic or juvenile forms, so that the Swiss-American anatomist Louis Agassiz (1807–73) 'having forgotten to ticket the embryo of some vertebrate animal . . . cannot now tell whether it be that of a mammal, bird or reptile'. Embryonic characters seemed consequently to be of greater value in determining the taxonomic position of an organism than its adult form, which could become greatly modified to accommodate to the contingencies of some idiosyncratic niche. Darwin's barnacles, which showed just such specialisations in adult life had not been recognised to be Crustacea until their less modified and obviously crustacean larvae had been identified. Why should this be so? If organisms were independently created, why should they not be as characteristic and unique in their earlier developmental stages as later on in life? Darwin recognised homology and the greater likeness of embryos with each other as further deducible consequences from evolution, since the structural differences which distinguish any two related forms must have been imposed by sequential slight modifications of a pre-existing form. A complex and competitive environment to which an organism is already minutely adapted will only allow very slight modification of existing structure. Although, as Darwin recognised, the variation in constitution which leads to a slight adjustment of adaptive structure or behaviour is present throughout the development of an individual, it does not become fully apparent until the stage of life at which it is adaptively relevant. For organisms whose early development is largely protected from variations in the environment, such as the mammals, adaptive variation will accumulate predominantly in adult forms but be progressively less apparent earlier and earlier in development.

However, if the imperatives of selection favour highly adaptive modification of early developmental stages to the external environment there is no necessary obstacle to its occurrence. As Darwin pointed out, in many organisms, larval stages are functionally equipped quite differently from the adult and serve entirely distinct roles in the life of the species, as with caterpillars and butterflies. What Darwin did not understand, except perhaps in the most general terms, was that development is a complex interactive process and is strongly resistant to variation, just as the adaptation of a species to its niche is complex and interactive and is, in general, intolerant of dramatic change. A new variation is thus subject to selection not merely in its effects on the adult structure, but even more so in its ability to permit organised embryonic development at all. Because the unravelling of form in embryonic life is arranged sequentially, the earlier in embryonic life a variation is expressed, the more drastic are its consequences for development. Embryos are thus inherently conservative and strategies of embryonic development are remarkably consistent within very large groups of organisms. The phenomenon of homology, or correlated structures, reflects this conservatism, and Darwin was certainly right in arguing that homological similarities indicated common ancestry.

Although the main argument of the *Origin of Species* can be presented easily enough, no such synopsis can display the intellectual intensity of the whole. The idea of a tree-like arrangement of organisms was common currency before Darwin. Darwin's development of this metaphor in genealogical terms illustrating the relationship between the variation and extinction of organisms united by the bond of inheritance through time will serve to summarise the main conclusions of the *Origin of Species*:

The affinities of all the beings of the same class have sometimes been represented by a great tree. I believe this simile
largely speaks the truth. The green and budding twigs may
represent existing species; and those produced during each
former year may represent the long succession of extinct species. At each period of growth all the growing twigs have
tried to branch out on all sides, and to overtop and kill the
surrounding twigs and branches, in the same manner as species and groups of species have tried to overmaster other species in the great battle for life. The limbs divided into great
branches, and these into lesser and lesser branches, were
themselves once, when the tree was small, budding twigs;
and this connexion of the former and present buds by ramify-
ing branches may well represent the classification of all extinct and living species in groups subordinate to groups. Of
the many twigs which flourished when the tree was a mere
bush, only two or three, now grown into great branches, yet
survive and bear all the other branches; so with the species
which lived during long-past geological periods, very few now
have living and modified descendants. From the first growth
of the tree, many a limb and branch has decayed and dropped
off; and these lost branches of various sizes may represent
those whole orders, families, and genera which have now no
living representatives, and which are known to us only from
having been found in a fossil state. As we here and there see a
thin straggling branch springing from a fork low down in a
tree, and which by some chance has been favoured and is still
alive on its summit, so we occasionally see an animal like
the Ornithorhynchus [Duck-Billed Platypus] or Lepidosiren
[lung-fish] which in some small degree connects by its affinities two large branches of life, and which has apparently been
saved from fatal competition by having inhabited a protected
station. As buds give rise by growth to fresh buds, and these,
if vigorous, branch out and overtop on all sides many a fee-
bler branch, so by generation I believe it has been with the
great Tree of Life, which fills with its dead and broken
branches the crust of the earth, and covers the surface with
its ever branching and beautiful ramifications. (O 129–30)

5 Sex, variation and heredity

Darwin was convinced from 1837 onwards that sexual reproduction or cross-fertilisation had an important place in the
theory of evolution. He formulated two major generalisations
concerning the sexual process. Firstly, that sexual reproduction,
unlike other modes of propagation such as simple division in
primitive organisms, or budding in many plants and lower
animals, was virtually a pre-requisite for individual variation.
Offspring of a sexual union always differed, however slightly,
from their parents, while non-sexual reproduction nearly always
produced only replicas. Secondly, sexual reproduction, unlike
other reproductive mechanisms, was a universal attribute of
animals and plants.

These two generalisations can be seen in the notebooks as
part of a deductive argument concerning the evolutionary process. Variation is a necessary condition of evolution by natural
selection, sexual reproduction is a necessary condition for variation, therefore sexual reproduction is a necessary condition for
evolution by natural selection. Since only those organisms exist
which can evolve, all organisms must engage in sexual reproduction.

Probably the argument never reached the *Origin* in this form
partly because Darwin was unable to discover why variation was
correlated with sexual reproduction. Furthermore, another and
seemingly antithetical attribute of sexual reproduction was that
it promoted the unity and constancy of species. By dispersing
individual variation in an interbreeding population, sex ensured
that the species unit could remain relatively constant, while re-
taining the opportunity to evolve at a rate appropriate to the
leisurely process of organic and geological change. Darwin saw
evolution as a process which affected a group of similarly but
not identically constituted individuals: strictly speaking it was
not the individual that evolved, but the group. Without sex
'there would be as many species as individuals'.

Beyond the recognition that sexual reproduction played some

key role in evolution, Darwin's contributions to sexual biology are extraordinary in their scope. It is not difficult to see why the subject was of such paramount interest: the reproductive process was the thread which connected all living things. But there was an additional aspect of sexual biology which was of great polemical value for Darwin. A creationist could account for the existence of most of the organised systems of plants and animals in terms of the purpose for which they had been created. The sexual apparatus and behaviour could be dealt with equally well in the same terms. But Darwin was able to ask a more basic question, what is sexual reproduction for, and why is sex the universal mode of propagation when other mechanisms were known to be possible? The only answer seemed to be that the 'purpose' of sexual reproduction was evolution.

Fortified by the conviction that sexual reproduction was, for some reason, a necessary condition for evolution, he investigated the phenomenon of hermaphroditism, with the expectation of finding that self-fertilisation, even if possible, would never exclude cross-fertilisation. Even though most plants and many lower animals were functional hermaphrodites, they invariably showed adaptations which enabled them to cross-fertilise when the opportunity arose. In a succession of major books Darwin explored the evolution of sexual mechanisms in flowering plants. More than anybody else, Darwin is responsible for the now obvious generalisation that flowers are devices to ensure cross-fertilisation by insects. In the amazing adaptations of orchid flowers to insect pollination, which he analysed in great detail, Darwin found the most striking evidence for the importance of cross-fertilisation among living things in general.

> Finally, if we consider how precious a substance pollen is, and what care has been bestowed on its elaboration and on the accessory parts in the Orchideae, . . . self-fertilization would have been an incomparably safer and easier process than the transportal of pollen from flower to flower . . . It is hardly an exaggeration to say that Nature tells us, in the most emphatic manner, that she abhors perpetual self-fertilization. (F 293)

In an attempt to provide a form of explanation for the need

for cross-fertilisation, Darwin began a long series of experiments on the effects of cross- and self-fertilisation in plants. He discovered that self-fertilisation often produced seeds of greatly diminished vitality, diminutive size and low fertility. Many plants, apparently physically equipped for self-fertilisation, were completely self-sterile. In some cases, such as the common primrose, he found that species were divided up into structurally distinctive groups which could freely cross-fertilise between groups but were nearly or completely sterile within a group. In fact, Darwin had difficulty resolving cause and effect in these phenomena. Superficially, it seemed reasonable to argue that cross-fertilisation had evolved as a consequence of the immediately disadvantageous effects of self-fertilisation, as indeed he did argue in several places, but he was obviously also aware that both self-sterility and cross-fertilisation probably had a common cause, the same 'final cause' which he dimly saw as being responsible for sexual reproduction in general.

Modern evolution studies still have great difficulty with the evolution of sexual mechanisms. Darwin did not solve, but rather created, the basic problem, which was, and still is, how to reconcile the obvious immediate advantage to an individual in self-fertilising, or producing young by some non-sexual mechanism like parthenogenesis, with the fact that this advantage is nearly always passed up in favour of a sexual process.

The theory of evolution by natural selection proposed that the details of organic structure were subject to the test of adaptiveness, but the only test of adaptiveness was success in reproduction. Reproduction was, therefore, literally an end in itself. In recognising this, Darwin also saw the importance of structures in sexually separated species of animal whose function seemed to be limited to mating success independent of other selective pressures. These structures were all those where 'the males and females of any animal have the same general habits of life, but differ in structure, colour, or ornament . . .' (O 89). In species with separated sexes the only *necessary* differences, seen from the point of view of natural selection, would be those connected directly with the exigences of the reproductive process itself. Yet clearly the differences between males and females in many species of animal far exceeded this necessary minimum.

Darwin explained the occurrence of such characters in terms of what he called 'sexual selection'. 'This depends, not on a struggle for existence, but on a struggle between the males for possession of the females; the result is not death to the unsuccessful competitor, but few or no offspring' (O 88), and the sexual differences arise because 'individual males have had, in successive generations, some slight advantage over other males, in their weapons, means of defence, or charms; and have transmitted these advantages to their male offspring' (O 89–90).

The case for sexual selection as an agent in the determination of structure or behaviour is supplementary to that for artificial selection, where man is the selector, and to that for natural selection, where the 'prevailing conditions' select. In sexual selection, only the males or females of the species select. It is also the extreme case of a principle which Darwin emphasised repeatedly in the *Origin*, that the physical conditions of life were less important than relations between organisms in determining the outcome of the struggle for existence. Since members of the opposite sex constituted the most evolutionarily significant resources in the environment, competition for reproduction should be extremely effective.

Sexual selection was a phenomenon of particular interest because sexuality was a privileged class of adaptation that could be referred only to the evolutionary process itself. Since selection operated on individuals, and on species only through their reproductive success or failure, it followed that sexual performance in all aspects would be subject to strict selection. The consequences of this selection for either appearance or behaviour could not necessarily be predicted by reference to any 'external' conditions of the kind that so obviously determined, for example, that carnivores should possess teeth and claws. Creationists were thus forced to argue that striking sexual characters were either the Creator's *jeux d'esprit* or were formed only for the delight of man. Darwin did not bother to deal directly with this argument because the theory of sexual selection was developed in full after the battle over the *Origin* had largely been won. But it is implicit in his insistence that to classify some character as beautiful and thereby self-justifying is to ignore the obvious impact that this 'beautiful' character has on other organisms of the same species.

Many of the phenomena of sexual selection seemed best described in terms of the appreciation of beauty by non-human animals. If the appreciation of beauty could lead to sexual adornment in animals, so a similar process could be argued to have occurred in man. The whole case for sexual selection is in fact an enormous appendage to Darwin's book, *The Descent of Man, and Selection in Relation to Sex* (1870), and forms an integral part of the larger argument that even man's most unique attributes find analogies among the beasts.

Inheritable variation is a necessary condition for evolution. Because organisms vary, selection is non-random, and because variation is inheritable, populations of organisms change as one generation succeeds another. In the mid-nineteenth century there was no scientific theory of variation, and Darwin was forced to search not so much for a mechanism or formal description of the principles of variation, but rather for precepts which would serve as working generalisations. It is probably one of Darwin's greatest achievements that he insisted throughout his life that variation itself could not act as a directing force in evolution. It was this claim which roused the physicist Sir John Herschel to castigate the *Origin* as 'the law of higgledy-piggledy', and which slammed the door on any personified designer and denied purpose in the evolutionary process.

Darwin's view of variation was enormously influenced by his belief that variation was something caused, rather than given. No organism that bred sexually was exempt from variation, but in his mature writings Darwin explicitly attempted to dispose of what he seems to have felt was an unscientific and virtually metaphysical view that variation was intrinsic to the reproductive process. As emphasised earlier in this chapter, the correlation between reproduction and variation was clear in his mind, but his view of the relationship may best be expressed by saying that the sexual process allowed the causes of variation to become manifest, rather than being itself the cause of variation.

Darwin's position was entirely consistent with his unyielding materialism in every manifestation of the organic world. If variations were causeless they would be outside the realm of scientific investigation; and Darwin never seriously considered allowing one of his axiomatic generalisations to be non-investigable, even if, for some purpose, it might be convenient to have causeless variation. In the absence of any knowledge of the physical basis of inheritance and variation Darwin invoked an active role for 'the external conditions of life' as an agent in the induction of variation. Exactly how he viewed this causative connection has been widely discussed, but it is clear that he saw heterogeneity in an organism's environment, no matter how subtle, as a necessary cause of variation:

> variability of every kind is directly or indirectly caused by changed conditions of life. Or, to put the case under another point of view, if it were possible to expose all the individuals of a species during many generations to absolutely uniform conditions of life, there would be no variability. (V ii. 255)

This is one of the few major generalisations that Darwin ventured which is now known to be categorically false. Variability is intrinsic in living things although it is not uncaused. However, the crucial issue surrounding Darwin's 'changed conditions' view was whether the relationship between the nature or direction of the change and the consequent variation was predictable or not. If changes in external conditions favoured adaptive variation, that is changes which could be seen to fit the organism better for its new conditions of life, then the interaction between the environment and individual development could be said to be constructive, and replaced natural selection as the principle cause of adaptive change. Such 'constructive' theories of variation had been central to the 'philosophical' evolutionism of Erasmus Darwin and of Lamarck. Alternatively, the influence of change of environment was merely to induce variations that were unpredictable in direction, in which case natural selection of accidentally adaptive variation was the cause of adaptive change. There can be no doubt that, for Darwin, the great weight of the evidence was in favour of an unpredictable effect of 'conditions' on variation, and for the pre-eminent role of natural selection in determining the direction of evolution. However, Darwin certainly came to admit the possibility of a constructive influence of the environment on variation, and included a new section on the inherited effects of habit and the use and disuse of organs in later editions of the *Origin*.

The inconsistency in Darwin's later writings on the nature of variation arose from his anxiety to provide an explanation for the relationship between heredity, that is the transmission of an organism's characters from one generation to another, and variation, that is 'errors' in transmission which, as emphasised, he saw as being caused by environmental change. In order to arrive at a relation of cause and effect between the action of the environment on a developing organism and the characters it transmitted to its offspring, Darwin proposed what he termed 'the provisional hypothesis of pangenesis'. The character of an organism's reproductive system, which in turn decides what it transmits to its progeny, was supposed to be determined by its ability to collect from the body-tissues particulate elements which Darwin called gemmules, each of which represented in a condensed or concealed form the differentiated character of the tissue from which it came. Development consisted of the orderly assembly and expression of the gemmules transmitted via the parental germ cells to the progeny. If the environment could affect development, for example, by enlarging an organ through extensive use, the number and also perhaps the quality of the gemmules sent to the reproductive system from that organ would be changed. Environmentally induced change in form would, therefore, tend to be inherited. Darwin was prepared to extend this argument even to behaviour, since he viewed behaviour as a manifestation of the physical organisation of an animal and therefore subject to the same rules as physical form. Learned behaviour could therefore become inherited and appear as instinct in later generations.

There are many defects in pangenesis. Most of the 'facts' used in its support, especially those on the inherited effects of use and disuse, were based on faulty observations, and its conceptual foundation was flawed by the general ignorance of his time about such basic processes as the nature of fertilisation and even about the cellular composition of tissues. The sex cells were a 'condensate' of the individual's gemmules which were transmitted through the process of fertilisation into a new individual whose characters represented a fusion of the characters of the two parents. The new individual carried 'hybrid' gemmules which governed the formation of the intermediate characters expressed by a hybrid. Darwin was, however, well aware that inheritance did not always entail blending of the characters of two parents: frequently the characters of one parent predominated. Sometimes apparently 'ancestral' characters would appear unexpectedly in a cross between two individuals known to have been separated for many generations from a common parent. This 'reversion' of characters was part of the evidence Darwin had marshalled to support the argument that even bizarrely modified and highly distinct breeds of domestic pigeon were all descended from the wild rock pigeon. Although Darwin did offer an extempore modification of the hypothesis to accommodate these phenomena, pangenesis was fundamentally a theory of 'blending' inheritance in which novel variation in one individual would tend to be eliminated by hybridisation with an individual of the normal type. It was thus strictly in conformity with Darwin's mature view that the sexual process was essentially conservative and tended to maintain the species character, rather than promote the persistence or spread of diversity.

In its two most fundamental attributes, pangenesis thus failed in the tasks that Darwin's earlier conceptions seem to have required of a theory of variation and inheritance. First, according to pangenesis, variation could only be due to an individual's adaptive response to an environmental stimulus to change. Secondly, it failed to account for any case of non-blending of characters in hybrids, or of reversion to ancestral characters. It was, therefore, even according to Darwin's own requirements, rather a bad hypothesis, and his most 'Darwinian' supporter, T. H. Huxley, disapproved of it strongly.

It would surely have been easier for Darwin's reputation had he not published pangenesis. By seeming to remove the accidental quality of variation, it diminished the constructive role of selection in adaptation and opened the way to a number of post-Darwinian attempts to accommodate evolution to a directed or determined process of change which let the Great Designer in again by the back door. But, however fond Darwin may have been of his pet theory (he referred to it once as 'my beloved child'), he still remained in favour of accidental variation whenever he thought generally about natural selection. Immediately after formulating the pangenesis hypothesis, Darwin used a metaphor that completely denies a constructive interaction between environment and variation.

Although each modification must have its proper exciting cause, and though each is subjected to law, yet we can so rarely trace the precise relation between cause and effect, that we are tempted to speak of variations as if they spontaneously arose. We may even call them accidental, but this must be only in the sense in which we say that a fragment of rock dropped from a height owes its shape to accident. (V ii. 420)

The important and original aspect of pangenesis was that it concentrated on the nature of the material bond between generations. It insisted that there must be such a bond, that its properties would clarify the unsolved problems of heredity and variation, and that only in such material terms could the mechanism of evolution be solved.

In the event, the fundamental mechanism of evolution has now been largely solved, and the solution has come, as Darwin guessed it would, from the study of the material bond between the generations. The attributes of the genetic material, DNA, are such that evolution by natural selection can now be seen to be a necessary property of matter organised in a certain kind of way. The famous double helix of DNA, solved by Watson and Crick in 1953, is a structure that can both carry hereditary information and replicate itself and the information it carries with extraordinary but not complete accuracy. It thus embodies in a pure chemical substance the three Darwinian conditions for evolution: heredity from the accuracy of self-replication, multiplication from the fact of self-replication, and variation from the rare inaccuracies of replication.

Precisely how genes are constructed and operate in living things, does not concern us here. In the context of Darwin's reflections on the nature of variation, however, it is interesting to see how nicely replication errors in DNA answer to the broken-rock metaphor quoted above. DNA carries hereditary information in the sense that the precise chemical structure of the DNA carried in sperm and egg determines the direction development will take in the growing organism. Unlike Darwin's gemmules, however, DNA does not represent in any sense a condensed version of the developed character. Rather, DNA carries chemically coded instructions for development, but the form of the code is connected quite arbitrarily with its 'meaning' for development, just as the form of a word is connected arbitrarily with its meaning. If a single letter in a word is changed at random, the meaning may be changed too, or the new word may have no meaning. So it is with DNA. The coded form of instructions for the developmental process is, like the written word, subject to misprints when it is reproduced. Selection operates through the reproductive success of an organism and not directly on its DNA. If a variant form of DNA resulting from some chemical accident in the coded instructions modifies a developmental process so that the reproductive performance of the variant organism is superior, then the new DNA instruction will gradually spread through the population and displace its inferior precursor. This might be called the gene's-eye view of evolution, in which organisms, the resultants of the developmental process, are seen as more or less efficient ways of replicating DNA.

Because DNA carries hereditary information in a chemically coded form, if we wish to determine the 'cause' of a variation it is useless to consult the decoded message, i.e. the developed structure of the organism itself. Because of the purely formal or arbitrary relationship between DNA instructions and the developmental process, environmental influences may change the outcome of development, but they cannot change the instructions coded in DNA in any predictable way that is adaptively related to the environment. There is no known or presently imaginable chemical route through which a developmental consequence of the interaction of an organism with the environment can cause a constructive, i.e. adaptive, change in the instructions coded in DNA.

It is hardly surprising that Darwin was unable to 'trace the precise relation between cause and effect' in the origin of variation. The separation of genetic information from the developmental process by the decoding system ensures that variations will appear as if they arose 'spontaneously' or 'accidentally', un-correlated with any visible inductive cause. Hereditary variation resulting from chemical instability in DNA must invariably pre-cede the test of adequacy in reproductive performance. The en-vironment of an organism plays a purely selective role in evolu-tion. If it appears to be constructive, it does so exclusively by selecting adaptive variants from the available repertoire.

The identification of DNA as the molecule in which the evo-lutionary obligations of organisms are invested makes a com-mon origin for all living things seem certain rather than possible or probable. Not only do all organisms from the most primitive bacteria to man use the same material to maintain the continuity of life, but they also use the same genetic code to translate hered-itary information into an adaptive developmental process. Since the genetic code is arbitrary and, so far as is known, abso-lutely unconstrained by chemical necessity, the only reasonable explanation for its universality is that it has evolved only once.

It is obvious that if evolution by natural selection is ultimate-ly guaranteed by the properties of a chemical substance, DNA, no special conceptual problem is raised by the origin of life. Darwin himself was prepared to admit that life could gradually emerge by comprehensible chemical processes from non-living precursors

if (and oh! what a big if!) we could conceive in some warm little pond, with all sorts of ammonia and phosphoric salts, light, heat, electricity, etc. present, that a proteine compound was chemically formed ready to undergo still more complex changes . . . (L iii. 18)

6 Man

It is sometimes said that Darwin hid the question of the evolution of man when he put the *Origin of Species* before the public. But the core of the argument in the *Origin of Species* was explicitly derived from Malthus's *Essay on Population*, which demonstrated the principle of geometrical increase in human populations. The continuity between man and the animals was obvious, and there was no polemical advantage in drawing attention to it further. Most references to man in the *Origin* deal with matters of inheritance, variation or structural homology. Darwin also used human behaviour to illustrate aspects of animal behaviour in general, but the larger implications of the theory of evolution for the origin of man's higher faculties he left with the minimal comment, 'Psychology will be based on a new foundation, that of the necessary acquirement of each mental power and capacity by gradation. Light will be thrown on the origin of man and his history' (O 488).

With the publication of T. H. Huxley's *Evidences as to Man's Place in Nature* in 1863, the elementary physical point of man's close taxonomic affinity with the higher apes was effectively settled, and in the same year Lyell's *Antiquity of Man* put beyond dispute the fact that man's origins lay in geological rather than historical time. In *The Descent of Man* (1871) Darwin summarised the points linking man to a general evolutionary scheme. Man was subject to heritable variation in innumerable slight characters, and his reproductive capacity normally outweighed the capacity of the environment to support his progeny, hence man must be subject to natural selection. Like many other species with an extended range, man presented geographically separated varietal forms or races. (He blamed a mixture of ignorance and self-interest for the common belief that the more distinct races of man were separable species: 'Has not the white man, who has debased his nature by making slave of his fellow Black, often wished to consider him as other animal'.) Physically, man was built on the same fundamental plan as other animals. Mentally, 'man and the lower animals do not differ in kind, although immensely in degree. A difference in degree, however great, does not justify us in placing man in a different kingdom . . .' (D i. 186).

Man, therefore, must be viewed as an animal species, unique and distinctive in a number of respects, but no more fundamentally unique or distinctive, even in intellectual powers, than the ants and the bees are when compared with other insects less spectacularly endowed with complex behaviours. Distinctiveness itself was an inevitable consequence of the evolutionary process. Extinction of intermediate precursor forms ensured that perfect continuity between species could never be observed, and the fossil record was unlikely to be of more evidential value for the origin of man than for any other individual species. The impressive collection of fossils that we now possess which may represent forms more or less directly on the line of human evolution from an extinct ape-like creature were unknown in 1871. Darwin thought they would be hard to discover because the evolution of man in his pre-human state would be localised, widespread dissemination of the species coming after the evolution of man's characteristic qualities. Using now as a predictive device the argument which previously had explained the proximity in space between recent fossils and living relatives, he guessed correctly that Africa would yield the earliest fossils ancestral to apes and men since man's closest living relatives, the gorilla and chimpanzee, were African species. Finally, from the lack of human remains in any but the most recent fossil strata, Darwin concluded that the evolution of man had been recent and exceptionally rapid, and that intermediate forms would have been quickly extinguished by their more successful cousins, thus extending the apparent gap between modern man and his closest living relatives.

Darwin's treatment of the process of 'humanisation' of the extinct ape-like progenitor of man was his most intensive imaginary reconstruction of an evolutionary trend, although many particular points in the structural evolution of man had been proposed by speculative evolutionists before him. Like all evolutionary reconstructions, it rests heavily on plausibility rather than evidence to command assent. Darwin's intention in this and other reconstructions of correlated evolutionary processes was precisely to indicate how it is possible to bridge seemingly impassable discontinuities of type by plausible intermediates. There are two questions to which an answer must be attempted to achieve a plausible evolutionary reconstruction; first, what was the route of transition, and secondly, why was that route followed? The answer to the latter question is often more conjectural than that to the former. It depends on discerning an advantage for a transitional modification, on imagining some crack in the surface of nature into which, by appropriate change, an evolving group can wedge itself. If, however, evolution by natural selection is accepted as the definitive and exclusive mechanism whereby organisms acquire their specific characteristics, a succession of transitions enforced by a succession of advantages is a certainty.

As soon as some ancient member in the great series of the Primates came, owing to a change in its manner of procuring subsistence, or to a change in the conditions of its native country, to live somewhat less on trees and more on the ground, its manner of progression would have been modified; and in this case it would have had to become either more strictly quadrupedal or bipedal . . . Man alone has become a biped; and we can, I think, partly see how he has come to assume his erect attitude, which forms one of the most conspicuous differences between him and his nearest allies. Man could not have attained his present dominant position in the world without the use of his hands which are so admirably adapted to act in obedience to his will . . . But the hands and arms could hardly have become perfect enough to have manufactured weapons, or to have hurled stones and spears with a true aim, as long as they were habitually used for locomotion . . . Such rough treatment would also have blunted the sense of touch, on which their delicate use largely depends. From these causes alone it would have been an advantage to man to have become a biped; but for many actions it is almost necessary that both arms and the whole upper part of the body should be free; and he must for this end stand firmly on his feet. To gain this great advantage, the feet have been rendered flat, and the great toe peculiarly modified, though this has entailed the loss of the power of prehension . . . If it be an advantage to man to have his hands and arms free and to stand firmly on his feet, of which there can be no doubt from his pre-eminent success in the battle of life, then I can see no reason why it should not have been advantageous to the progenitors of man to have become more and more erect or bipedal. (D i. 140–2)

Darwin continues this plausible reconstruction through to the effect of upright posture and increased brain size on the structure of the skull, shape of the face, dentition, flexure of the spine, broadening of the pelvis, and so on.

It is important to understand that reconstruction arguments of this type do not imply the guidance of an imperious external necessity which transcends the immediate advantage. Alfred Russel Wallace, the naturalist who independently identified the principle of evolution by natural selection, was unable to accept that the definitive stages in the evolution of man could have been due solely to the exigences of a natural environment. Wallace's position was that man was more perfect than he needed to be, particularly in mind, for any conceivable requirement of natural selection, and that 'a superior intelligence has guided the development of man in a definite direction, and for a special purpose, just as man guides the development of many animal and vegetable forms.'

Darwin's response to the introduction of this novel agency was to insist sternly on the absence of any limit to the action of selection by comprehensible, if not definitely known, causes. Man evidently owed to his intellectual attributes the physical and cultural inventions which had made him

the most dominant animal that has ever appeared on the earth . . . These several inventions, by which man in the rudest state has become so pre-eminent, are the direct result of the development of his powers of observation, memory, curiosity, imagination, and reason. I cannot, therefore, understand how it is that Mr. Wallace maintains that 'natural selection could only have endowed the savage with a brain a little superior to that of an ape'. (D i. 136–8)

The evolution of man's higher intellectual faculties raised both semantic and metaphysical problems, but Darwin found that the comparative method which dealt successfully with less contentious but none the less complex evolutionary processes could under certain conditions be applied to mind as well. He realised as early as 1838 that to concentrate on the subjective aspects of human thought would be to restrict discussion entirely to a human context and to exclude as unknowable any equivalent in animals. 'The centre is everywhere and the circumference nowhere as long as this is so –!!' (T ii. 109). The search for the 'circumference' meant an examination of animal behaviour, and a re-description of human behaviour in terms of animal analogies: 'Origin of man now proved – Metaphysics must flourish – He who understands baboon would do more toward metaphysics than Locke' (M 281).

The enterprise involved exploring the reasonable extensions of language commonly used for states of the human mind to illuminate animal behaviour which seemed to be contextually analogous to the human actions by which such mental states were normally accompanied. In so far as these extensions seemed plausible, they constituted evidence in favour of the continuity of faculty between animals and man. If it was reasonable to say that a dog behaved as if it was 'fearful' when threatened with a stick, then fearfulness might legitimately be extended to dogs. Such an extension of usage points only to a similarity of behaviour in context between animals and man, and is logically similar to arguing that a cringing human, similarly threatened, whose state of mind we do not inspect through language, is fearful. 'Forget the use of language and judge only by what you see' (M 296). Darwin's comparative metaphysics dealt with the problem of other minds by pure and literal anthropomorphism. Indeed, 'Having proved mens and brutes bodies are one type: almost superfluous to consider minds' (T iv. 163). Darwin viewed metaphysical objections to extending human mental qualities to the animals as 'arrogance'.

Various attempts have been made to attack Darwin for his

anthropomorphism and, concomitantly, to defend it as a mere convenient metaphor. Both the attack and the defence are certainly inappropriate. The whole point of Darwin's position was to indicate homological resemblance between human and animal behaviour, and it followed that it was no more absurd to speak of a higher mammal showing fear, reasoning power or pleasure than to call the structure on the end of a chimpanzee's forelimb a hand. The difference was one of degree not of kind. There was one continuous 'thinking principle' throughout the animals which Darwin viewed as being contingent on the presence of an organised nervous system, and consequently 'The difference between intellect of man and animals not so great as between living thing without thought (plants) and living thing with thought (animals)' (T i. 66)

The 'thinking principle' appeared in a number of manifestations throughout the animal kingdom. Some, like instinctive behaviour, seemed peculiarly applicable to animals, others, like reasoning power or conscience, peculiarly applicable to man. But one of Darwin's achievements was to show how blurred are the edges of such concepts:

> I will not attempt any definition of instinct. It would be easy to show that several distinct mental actions are commonly embraced by this term; but every one understands what is meant, when it is said that instinct impels the cuckoo to migrate and to lay her eggs in other birds' nests. An action, which we ourselves should require experience to enable us to perform, when performed by an animal, more especially by a very young one, without any experience, and when performed by many individuals in the same way, without their knowing for what purpose it is performed, is usually said to be instinctive. But I could show that none of these characters of instinct are universal. A little dose, as Pierre Huber expresses it, of judgement or reason, often comes into play, even in animals very low in the scale of nature. (O 207–8)

In higher animals the use of the term 'instinct' to describe complex behaviour became progressively more difficult because of the interference of increasingly large doses of judgement and reason:

The orang in the Eastern islands, and the chimpanzee in Africa, build platforms on which they sleep; and as both species follow the same habit, it might be argued that this was due to instinct, but we cannot feel sure that it is not the result of both animals having similar wants and possessing similar powers of reasoning. (D i. 36)

In man the term 'instinct' was most readily applied to the behaviour of babies and to those involuntary actions which accompanied particularly vivid sensations and emotions. The suckling of a new born baby, a snarl of rage or a cry of fear could easily be agreed upon as instinctive behaviour in man, and behaviour closely similar both in detail and in context could be found among the higher animals. In *The Expression of the Emotions in Man and Animals* (1873) Darwin pointed out that highly socialised birds and mammals used a variety of expressive modes to indicate fear, pleasure, arousal, alarm and affection. In mammals, including man, the face was commonly used as in wrinkling the brows or snarling, and it was often ornamented to give additional emphasis and visibility to such signs. In concentrating on infant behaviour and on the generality of expressive behaviour between humans in the savage and cultivated states, the innate could be broadly distinguished from the merely conventional features of expression. The presence of universal expressive forms in mankind as in animals emphasised the physiological basis of expression and reduced the special significance of 'private' mental agencies in guiding at least some features of man's relationship with man. Finally, perhaps most importantly, Darwin pointed to the serviceability of simple stylised expressions. They did not merely happen, they gave information to one individual about the mental state of another and had consequences not only for the expressor but for the individual towards whom the expression was directed.

The most distinctive feature of Darwin's human biology, and the consideration which enabled him to avoid the enfeebling invocation of a mysterious power to which Wallace and others felt obliged to resort, was his recognition of the overwhelming importance of socialisation in human evolution. This is the thread which unites the two seemingly disparate themes – the origin and nature of man, and sexual selection – in *The Descent of Man*, since the sexual organisation of a species was a necessary component of its social organisation. The behaviour which accompanied the interactions between parent and child, between male and female and between male and male were as much species characters subject to evolution by differential reproductive success as characters of a non-social kind, either physical or behavioural. In *The Descent of Man*, sexual competition and sexual choice were invoked to explain some of the physical attributes of man that did not seem to contribute directly to general biological advantage. The general lack of body hair compared with man's ape-like relatives and its different distribution in males and females Darwin attributed to sexual preference. Marked racial differences often appeared in sexually differentiated characters such as body hair distribution or relative height of male and female. He concluded that sexual preferences and the forms considered beautiful or ideal in the opposite sex were capricious within the human species but persistent enough in a given group to have induced physical modification of a racial character during the dispersal of primitive man over the continents.

In a period of growing awareness of the heterogeneity of customary behaviours in mankind, Darwin did not extend his relatively limited claims about the role of sexual behaviour in the creation of races to empty speculations about biological necessity determining the details of tribal, social or family structures. It was more important to attempt a plausible account of the evolution of such complex mental faculties as language, genius, the sense of beauty and the moral and religious senses which seemed to set the human species apart. He saw the behaviour and language associated with such ideas as serviceable functions, highly developed in humans, but having roots in the behaviour of pre-human animals. Indeed, the evolution of language through dialectal variation, its continuity, spatial heterogeneity, modification of ancient forms, rudimentary vestiges, and its adaptation to function in communication, were prevalent ideas in the early nineteenth century. Darwin used the evolution of language several times in *The Origin of Species* as an illustrative metaphor for the process of biological evolution.

In *The Descent of Man*, Darwin naturally sided with the phi-
lologists who looked to an origin for language in the primitive
expressive function of the voice:

> I cannot doubt that language owes its origin to the imitation
> and modification, aided by signs and gestures, of various
> natural sounds, the voices of other animals, and man's own
> instinctive cries . . . we may conclude from a widely-spread
> analogy that this power would have been especially exerted
> during the courtship of the sexes, serving to express various
> emotions, as love, jealousy, triumph, and serving as a chal-
> lenge to their rivals. The imitation by articulate sounds of
> musical cries might have given rise to words expressive of
> various complex emotions. (D i. 56)

This passage includes the clear implication that language, the
supremely social attribute of man, originated in the sociality imposed by sexual relations.

Darwin connected man's facility for language with his extraordinary reasoning power, or intellect. He regarded the ability to form novel mental associations as a more complex alterna-
tive to the relatively inflexible and determinate behaviour of instinct. Man's truly instinctive behaviours were apparently fewer
and simpler than those of his relatives. The human faculty of
language was an innate and extremely specialised adaptability
related to a generalised high level of intellectual development,
and clearly distinct from and more subtle than an instinctive
capacity to use a particular language: 'We must believe that it
requires a far higher and far more complicated organisation to
learn Greek, than to have it handed down as an instinct'
(M 339).

Darwin's view of language as a specialised adaptability is of
considerable interest. He expressed this notion as 'an instinctive
tendency to speak' which he inferred from 'the babble of our
young children'. The key point is that it was not language but
the tendency to language which was the distinguishing attribute. Darwin's failure to recognise adaptability in physical form
as a character subject to selection in its own right was largely
responsible for his confusion over the inheritance of acquired

characters. But in the matter of behaviour in general, and language in particular, the concept of adaptability was securely grasped.

Through his account of the development of language use, Darwin partially answered Wallace's objection that comprehensible selective agencies could account for only a slight increase in brain function over that of the apes:

> A great stride in the development of the intellect will have followed, as soon as, through a previous considerable advance, the half-art and half-instinct of language came into use; for the continued use of language will have reacted on the brain, and produced an inherited effect; and this again will have reacted on the improvement of language. The large size of the brain in man, in comparison with that of the lower animals, relatively to the size of their bodies, may be attributed in chief part . . . to the early use of some simple form of language, – that wonderful engine which affixes signs to all sorts of objects and qualities, and excites trains of thought which would never arise from the mere impression of the senses, and if they did arise could not be followed out. (D ii. 390–1)

It is essential to realise that when Darwin refers to an 'advance' or a 'development,' as he frequently does in his writings on man, it is always implied that this advance or development was achieved at the expense of the less advanced or developed. The environmental pressure which Wallace and others like him ignored was the competitive interaction between evolving lines of infra-humans, and not some inanimate influence such as climate, or animate influences of an incomparably lower grade such as predators or food resources.

Some level of socialisation is obviously a necessary condition for the evolution of language. Darwin went on to develop the argument that the origins of the moral sense could be found in the development and maintenance of social structure. As a matter of common observation, many animal species could be found to have developed a level of sociality. In social species certain adaptive consequences of social life could be identified, characterised generally as mutual aid. It therefore seemed reasonable that behavioural adjustments to social living could be developed by the normal processes of evolution. Leaving aside the idiosyncratic behaviour of particular animal societies, Darwin concentrated on the nature of the primary 'tendency to associate'. His view was that social animals take 'pleasure' in their association, and that this source of pleasure was an impulsion to behave in a sociable way. Subjectively, the anticipation of pleasure served as an impulse to action in man. Darwin clearly saw pleasure only as a concomitant of actions whose effects were biologically adaptive. It was a consequence of one adaptive action and a cause of its repetition, and depended on the biological value of the action itself. The successful expression of an innate tendency or instinct in the lower animals or in man was inherently pleasurable, but the sources of pleasure were fixed by evolutionary contingencies.

Darwin argued that the moral sense or conscience was derived from a feeling of dissatisfaction at the non-fulfilment of an innate tendency to sociable behaviour. As a consequence, the particular form taken by the moral sense, i.e. those specific actions which might in a particular society come to acquire the attributes of good or evil and earn the subjective judgement of conscience, was determined solely by the contingencies which led to a particular form of socialisation in that society.

Embedded in Darwin's discussion of the evolution of conscience out of social behaviour there is a profound evolutionary paradox which is still with us today. The paradox is that when an animal behaves in a 'social' way it will tend to sacrifice its individual advantage to the advantage of the group, and, as Darwin had pointed out in the *Origin*, natural selection should evolve selfish, not altruistic organisms. The extraordinarily 'unselfish' behaviour of sterile worker bees had led him not just to see the paradox of social behaviour, but also the form of the solution. The point was that each hive was really an extended family, and because of the principle of inheritance, natural selection could act as well through a family as through an individual.

Thus I believe it has been with social insects: a slight modification of structure or instinct, correlated with the sterile condition of certain members of the community, has been advantageous to the community; consequently the fertile males and females of the same community flourished, and transmitted to their fertile offspring a tendency to produce sterile members having the same modification. (O 238)

When dealing with the evolution of social behaviour in general, therefore, Darwin was convinced that it could only originate in extended family groups containing many closely related individuals, and therefore the early evolution of the human species must have been in such groups. No doubt for this reason, when he wished to illustrate the idea that particular moralities arise from particular contingencies, he turned to the hive bees for his example because here the conflict between individual and social imperatives was so easy to resolve in favour of the social.

In the same manner as various animals have some sense of beauty, though they admire widely different objects, so they might have a sense of right and wrong, though led by it to follow widely different lines of conduct. If, for instance, to take an extreme case, men were reared under precisely the same conditions as hive-bees, there can hardly be a doubt that our unmarried females would, like the worker-bees, think it a sacred duty to kill their brothers, and mothers would strive to kill their fertile daughters; and no one would think of interfering. Nevertheless the bee, or any other social animal, would in our supposed case gain, as it appears to me, some feeling of right and wrong, or a conscience . . . In this case an inward monitor would tell the animal that it would have been better to have followed the one impulse rather than the other. The one course ought to have been followed: the one would have been right and the other wrong. (D i. 73–4)

It is quite clear that Darwin realised how radical was his extension of previously restricted 'human' moral terminology to cover behaviour in non-humans.

The imperious word *ought* seems merely to imply the consciousness of the existence of a persistent instinct, either innate or partly acquired, serving him [man] as a guide, though liable to be disobeyed. We hardly use the word *ought* in a metaphorical sense, when we say hounds ought to hunt, pointers to point, and retrievers to retrieve their game. If they fail thus to act, they fail in their duty and act wrongly. (D i. 92)

From moral relativism to denying the unique moral authority of God was a short step. Darwin accepted Auguste Comte's view that the idea of potent spiritual agencies was a conception to which the role of cause for unexplained effects was attributed by man in a primitive condition. But although 'with the more civilised races, the conviction of the existence of an all-seeing Deity has had a potent influence on the advancement of morality', the evidence of anthropology spoke against this conviction having a basis fundamentally different from belief 'in the existence of many cruel and malignant spirits, possessing only a little more power than man; for the belief in them is far more general than of a beneficent Deity' (D ii. 394, 395).

For Darwin, the moralistic God of the civilised world was simply the personification of habitual convictions with origins in the social instincts of lower animals. There can be little doubt that Darwin was speaking of his own private morality when he wrote that the fully conscientious person was one who could say, 'I am the supreme judge of my own conduct, and in the words of Kant, I will not in my own person violate the dignity of humanity' (D i. 86).

Darwin did not develop a distinctive ethical philosophy from his naturalistic conception of the origin of the moral sense. His obscure but strangely intimate appeal to Kant presumably shows only that he had identified Kant's private categorical imperative to moral duty with a biologically based instinct to behave in a sociable way. To deny unique moral authority to a deity was certainly not equivalent to denying the moral value of actions which were promoted by religious teaching. Since moral worth was defined for Darwin by the general character of society, he felt free to applaud actions that he valued or criticise actions that he found morally repugnant, within the general moral framework of his own society.

Darwin was inconsistent about the source of moral improvement in mankind, though, true to the values of the liberal middle class, he did not doubt that such an improvement had occurred and that he was living in a by and large morally advanced condition. He simultaneously argued that by appropriate marriage man 'might by selection do something not only for the bodily constitution and frame of his offspring, but for their intellectual and moral qualities' (D ii. 403), and that 'the moral qualities are advanced, either directly or indirectly, much more through the effects of habit, the reasoning powers, instruction, religion, etc., than through natural selection . . .' (D ii. 404). In support of the former view he agreed with his cousin Francis Galton, the first human geneticist, that 'if the prudent avoid marriage, whilst the reckless marry, the inferior members will tend to supplant the better members of society' (D ii. 403), and that there should be 'open competition for all men; and the most able should not be prevented by laws or customs from succeeding best and rearing the largest number of offspring' (D ii. 403).

Probably the angriest debate in modern social science is over the extent to which man's highest social and intellectual qualities are governed by inalienable biological imperatives. Evolutionary sociobiology, which takes its principles from study of the evolution of animal societies, has moved so strongly towards a prescriptive application of these principles to man that Edward Wilson, a leading modern exponent of this neo-Darwinian cult, is now prepared to endorse as attainable and desirable objectives, both 'a culture predesigned for happiness' and 'a genetically accurate and hence completely fair code of ethics'.

It is clear that such ideas can find roots in Darwin's own writings. If the social constitution of mankind is not merely permitted by our genes but precisely specified, then the argument for a biologically based prescription for human moral behaviour is strictly identical to Darwin's claim that pointer dogs 'ought' to point because they are constitutionally adapted to do so. Furthermore, still following Darwin, pointers will be happier if they point. But Darwin was characteristically cautious in specifying what human behavioural attributes were. Just as the human faculty of language was best described as a generalised tendency to speak rather than as a specific endowment to speak

Greek, so the heterogeneity of other forms of sociality showed that these too were tendencies to adopt general classes of behaviour and were not precisely determined by inheritance. It seems likely that the immense cultural adaptability of our species will ultimately provide its own defence against the arrogant scientism of evolutionary sociobiology in its prescriptive application to mankind.

7 Perfection and progress

The concept of perfection in the natural world was integral to pre-Darwinian rationalisations of nature as evidence of the designing hand of a providential creator. The concept of progress had roots in speculative evolutionism as an inference from embryonic development and from the heterogeneity of animal and plant life in general, from the simplest organisms to man. In human affairs, idealisations of perfection and progress represented two opposed justificatory principles whose confrontation was a major feature of eighteenth- and early nineteenth-century thought. Anglican orthodoxy and conservative political principles insisted on the value of existing institutions, while revolution, romanticism and the rise of democracy insisted on the necessity of progress. To Darwin, idealisation of human institutions, whether static or in flux, was of marginal interest, but the philosophical extension of perfection and progress into the organic world as a source of explanatory principles was of great concern. Perfection and progress were abstractions which had no place in Darwin's pragmatic and relativistic scheme. The natural world must deviate from any reasonable definition of perfection as a universal quality because variation and selection necessarily implied heterogeneity in fitness. Extinction and death, seen not as divinely ordained but as a necessary effect of the passage of time on a population of organisms, ensured that 'perfection' in biological organisation could be defined only in relation to the environment in which an animal or plant lived.

Structural adaptation was the central phenomenon that both the Divine Artificer and natural selection were called on to explain. Darwin's dialectical position was based on the elementary proposition that not everything was as perfect as it might be, taking perfection in the broadest sense. If the Creator had the capacity to create organisms adapted for any context, why did his creative powers fail in oceanic islands, like New Zealand, where no mammals were to be found? Why was there often an incomplete correlation between habit and structure, as in the woodpecker of the South American plains 'where not a tree grows . . . which in every essential part of its organisation told me plainly of its close blood relationship to our common species; yet it is a woodpecker which never climbs a tree!' (O 184)? For Darwin, adaptation was simply the consequence of an interaction between contingency and time. If the contingency was too strict, the time too short or the competition too intense, extinction was the outcome.

The extreme cases of structural non-adaptation that had no meaning in terms of a perfect creation were rudiments, such as the traces of hind limbs and pelvis entirely enclosed in the body of a snake or a whale, or wings permanently enclosed by hard cases on the backs of beetles living on windy islands. For Darwin, these were relics of an organism's past that had been reduced in response to changed contingencies through evolutionary time to the point where, while still betraying their homologies, they served no purpose in the life of the organism.

> In reflecting on them, every one must be struck with astonishment: for the same reasoning power which tells us plainly that most parts and organs are exquisitely adapted for certain purposes, tells us with equal plainness that these rudimentary or atrophied organs, are imperfect and useless. In works on natural history rudimentary organs are generally said to have been created 'for the sake of symmetry', or in order 'to complete the scheme of nature'; but this seems to me no explanation, merely a restatement of the fact. Would it be thought sufficient to say that because planets revolve in elliptic courses round the sun, satellites follow the same course round the planets, for the sake of symmetry, and to complete the scheme of nature? (O 453)

Just as organic structure defied characterisation in terms of perfection except in the purely relative context of contingency, so animal behaviour too fell far short of perfection. A benign and providential Creator should have seen to it that the 'smell of man would be disagreeable to mosquitoes' (T ii. 103). It is obvious from Darwin's view of the origin of moral behaviour that contingency was the sole relevant determinant. The 'imperious word *ought*' could be defined only in context. In his *Trans-mutation Notebooks*, Darwin reduced the definition of biological perfection to the only sense that it could stand in his scheme, and, indeed, the only sense that it can stand today: 'perfection consists in being able to reproduce' (T vi. 159).

To the extent that the natural world was neither morally nor structurally perfect the evidence was clearly on Darwin's side. Nevertheless, the organic world was often so extraordinarily striking in the refinement and complexity of its adaptations that Darwin rightly foresaw difficulties in their acceptance as the outcome of mere variation and natural selection. Always invoking the 'great principle of gradation', Darwin's position was consistent. For complex or wonderful structures, the problem was one of degree, not of kind.

> To suppose that the eye, with all its inimitable contrivances for adjusting the focus to different distances, for admitting different amounts of light, and for the correction of spherical and chromatic aberration, could have been formed by natural selection, seems, I freely confess, absurd in the highest possible degree. Yet reason tells me, that if numerous gradations from a perfect and complex eye to one very imperfect and simple, each grade being useful to its possessor, can be shown to exist; if further, the eye does vary ever so slightly, and the variations be inherited, which is certainly the case; and if any variation or modification in the organ be ever useful to an animal under changing conditions of life, then the difficulty in believing that a perfect and complex eye could be formed by natural selection, though insuperable by our imagination, can hardly be considered real. (O 186–7)

Furthermore wonderful adaptations, such as wings for flight, were often not unique, as in birds and bats. If the Creator was in the business of ordaining flight he must have done so several times, solving the single problem of flying in entirely distinctive ways. Independently evolving lines may arrive at functionally similar solutions from different starting points and following different routes. The very fact of the non-uniqueness of a particular solution to a complex problem qualifies the whole idea of 'perfection'.

Many of Darwin's botanical works, and especially the *Fertilization of Orchids* (1862), *Climbing Plants* (1865) and *Insectivorous Plants* (1875) were further contributions to the destruction of the idea of nature as a collection of perfectly and discontinuously conceived contrivances for the ideal performance of certain tasks. The theme of Darwin's analysis of all these remarkable adaptations is the same. Descent with modification demands that in related groups of organisms there will be structural resemblances, while the conditions of life determine how in detail this hereditary endowment is turned to the service of the species. If any adaptation, no matter how curious or specialised, is looked at with an eye informed by the expectation of seeing structural conformity with related types, the uniqueness of the adaptation can be seen to be only partial. So the astonishing 'perfection' of the multiform variations in the flowers of orchids which ensure their fertilisation by particular insects, often themselves co-adapted to the orchid flower to an almost equally astonishing degree, could be resolved into a number of gradual modifications from some fundamental homological structures found in much less remarkable manifestations in more ordinary flowers. If specialised adaptations were the endowments of an omniscient Creator there could be no rationalisation for these subtly modified but nevertheless apparent feet of clay. If, however, opportunistic modification of pre-existing parts to meet contingency was the method of arriving at adapted structure, the underlying primitive order common to highly specialised and simpler flowers was explained. As a recent commentator, Michael Ghiselin, has memorably written, all such specialised adaptations are 'contraptions' rather than 'contrivances', evolved out of any machinery that was to hand.

For Darwin the whole concept of perfection was at best a useless and at worst a pernicious notion applied arbitrarily to those adaptations that seemed to conform best to a human view of how some biological task should be done. If the biological task was simply 'the ability to reproduce', no adaptation, however distasteful or seemingly incompetent, that permitted it could legitimately be denied the attribute of perfection.

If we admire the truly wonderful power of scent by which

the males of many insects find their females, can we admire the production for this single purpose of thousands of drones, which are utterly useless to the community for any other end, and which are ultimately slaughtered by their industrious and sterile sisters? ... If we admire the several ingenious contrivances, by which the flowers of the orchid and of many other plants are fertilized through insect agency, can we consider as equally perfect the elaboration by our fir-trees of dense clouds of pollen, in order that a few granules may be wafted by a chance breeze on to the ovules? (O 202–3)

The answer to Darwin's rhetorical questions was, of course, that our opinion is irrelevant. Either way, the animals and plants acted sufficiently well. The concept of perfection was thus replaced by that of mere sufficiency, and if sufficiency was achieved at a terrible cost it was not the less sufficient. But if natural selection imposed a cost, did it not also confer a benefit in terms of progress? Imperfections in nature were visible long before Malthus and failure of the persistent struggle to justify God's actions to man led through the eighteenth century away from a necessary stasis in the natural order to the more promising hypothesis of a necessary progress. By the middle of the nineteenth century virtually all philosophical speculation about the origins of things was evolutionary (though not Darwinian) in character. Nature was personified as an immaterial agency striving for ever more complex forms of being. The variety of living organisms represented stages in the expression of systematic progress towards a variously defined state of perfection, either infinitely elusive and unknowable or more or less clearly identified with the traditional virtues of the higher beings of Christian mythology. The human species represented the highest point yet reached in this inexorable ascent, and the wonders of the human mind in both its private and social manifestations pointed the way ahead.

For Darwin, the idea of a necessary progressive evolution was another gratuitous superimposition of human values on a fundamentally value-free process. A one-dimensional progressive law denied the biological equivalence of living species that was guaranteed by their persistent powers of reproduction. 'It is absurd to talk of one animal being higher than another – *we* consider those, where the cerebral structure/intellectual faculties most developed, as highest. – A bee doubtless would where the instincts were most developed' (T i. 50). And the development of mind, whether of the 'intellectual' or 'instinctive' type, was obviously not the only way of ranking living things since it left out the vegetable world, and 'who with the face of the earth covered with the most beautiful savannahs and forests dare to say that intellectuality is only aim in this world?' (T i. 72). Nevertheless, something which might reasonably be called progress had clearly happened in the biological world, although in many different dimensions. Exactly what this progression consisted in was not clear, but Darwin was inclined to accept the notion of complexity. Yet he was constitutionally sceptical about the necessity of the process. If complexity had tended to increase in the biological world, then this was a phenomenon that should receive an explanation that was not entirely self-referring. Again the crucial idea is the blindness of the evolutionary process, which responds only to contingency. Darwin realised that in general the contingencies favoured diversification: the very existence of one group of organisms created new niches for another group (as in the simple case of plants creating *ipso facto* the possibility of animals) and so on indefinitely. The level of complexity of organisms was thus a function of the level of complexity of the organic environment. Darwin was 'greatly interested' in the elegant idea (suggested to him in 1877) that the major evolution of the flowering plants, which was known from the fossil record to have been extremely rapid, had been under the 'impulse' of the evolution of flower-frequenting insects. Insects guaranteed rapid and efficient cross-fertilisation; plants lured them with flowers that were conspicuous or scented, and contained nectar. The two groups co-evolved with great rapidity. But, given appropriate circumstances, the evolutionary process should be, if not strictly reversible, at least retrograde.

Oceanic islands offered, for higher organisms, opportunities for retrograde evolution that were not available on the mainlands where their structural complexity had initially evolved. The flightless birds and insects of such islands had clearly lost a highly complex function. Similarly, blind animals that lived in caves or underground had sacrificed their sight to the contingencies of their new environment. In discussing this form of retrograde change, Darwin introduced a novel idea which brought him to the extremely modern statement that evolutionary change in general tends towards a maximum economy in the use of resources, a principle which follows directly from the Malthusian paradox that reproductive rates tend to exceed resource availability. Useless structures such as eyes in cave-dwelling animals were a waste of resources that could be traded off against reproductive performance. In such animals,

> The principle...of economy...by which the materials forming any part or structure, if not useful to the possessor, will be saved as far as is possible, will probably often come into play; and this will tend to cause the entire obliteration of a rudimentary organ. (O 455)

The power of the principle of economy in tuning the performance of an organism to the maximum capacity of its environment was demonstrated in Darwin's account of the formation of the hexagonal wax chambers that form the comb of the hive-bee. Darwin showed by an exemplary mathematical argument that the structure of the comb was precisely that which would minimise the amount of wax used by the swarm. Bees convert nectar, gathered from flowers, to wax for constructing the comb. They also convert nectar into honey and fill the comb with it. The honey is then used to provide a source of food for the swarm throughout the winter. If nectar is a limiting resource it is obviously of advantage to minimise the amount consumed in forming the comb and hence to maximise the amount available for forming the honey to store in it.

Thus the principle of economy served equally to account for both retrogressive and progressive evolution, for the destruction of one of nature's most 'perfect' structures, the vertebrate eye, and for the development of one of the most subtle examples of instinctive behaviour.

Evolution thus 'blindly' follows the route of maximum resource use. The route may be 'forward' or 'backward' depending on the observer's opinion. In general, the route would seem to be forward because competitive advantage lay always with the more efficient members of a species: the less efficient option being, as it were, already occupied for the time being by the other members of the species.

This view of evolutionary progress denied any necessary 'progressive tendency'. The human species was thus an accident of the evolutionary history of the earth. Man's most valued qualities were not, in principle, unique any more than a bird's wing was the unique solution to the problem of flight. 'What a chance it has been . . . that has made a man – any monkey probably might, with such chances be made intellectual, but almost certainly not made into man' (T iv. 166).

Darwin's solution to the problem of progressive evolution was rejected by many of his contemporaries because it disallowed providential design. After evolution became generally acceptable, providence could still be seen as the guiding hand. If variation was the raw material on which natural selection acted, then providential intervention could be invoked to account for the progressive tendency in evolution, and in particular for the purposive origin of man, by providing favourable variants at each stage in evolution. In rejecting this argument, Darwin finally excluded all transcendental agency from the process of evolution. His rejection was based on the correct conviction that variation and the 'use' to which it is put were independent of one another. Variations put to use by man in domesticated species were the analogy, but the argument was generalised. Did an omniscient Creator really foresee every variation which would enable any organism, either through human caprice or in the wild, to outbreed its cousins?

> Did He ordain that the crop and tail-feathers of the pigeon should vary in order that the fancier might make his grotesque pouter and fantail breeds? Did He cause the frame and mental qualities of the dog to vary in order that a breed might be formed of indomitable ferocity, with jaws fitted to pin down the bull for man's brutal sport? But if we give up the principle in one case . . . no shadow of reason can be assigned for the belief that variations, alike in nature and the result of the same general laws, which have been the groundwork

through natural selection of the formation of the most perfectly adapted animals in the world, man included, were intentionally and specially guided. However much we may wish it, we can hardly follow Professor Asa Gray in his belief 'that variation has been led along certain beneficial lines', like a stream 'along definite and useful lines of irrigation'. If we assume that each particular variation was from the beginning of all time preordained, the plasticity of organisation, which leads to many injurious deviations of structure, as well as that redundant power of reproduction which inevitably leads to a struggle for existence, and, as a consequence, to the natural selection or survival of the fittest, must appear to us superfluous laws of nature. On the other hand, an omnipotent and omniscient Creator ordains everything and foresees everything. Thus we are brought face to face with a difficulty as insoluble as is that of free will and predestination. (V ii. 431–2)

8 Darwinism and ideology

In 1980 the two leading contenders for the most influential secular office on earth, the Presidency of the United States of America, were rivals also in their anxiety to declare publicly their belief in the literal truth of the biblical story of creation. This was a sobering reminder that Darwin's theory of evolution has not commanded the universal assent to which its early adherents felt it was entitled. It is extremely galling for a biologist to realise that almost exactly a hundred years ago T. H. Huxley was wrestling over precisely the same issue with Mr Gladstone, the once and future Prime Minister of Great Britain. Indeed, we must fear that the situation will get worse as the gulf between scientific understanding and popular belief inexorably widens. For it is unfortunately true that the final clarification of those points which Darwin struggled hardest and least successfully to solve has come from the new science of genetics, and although the properties of genes explain much of organic evolution with gratifying simplicity, genes do not recommend themselves to anybody unaccustomed to conceiving the real existence of objects smaller than can be seen with the naked eye.

Although Darwin's arguments implied that things with the properties of genes would eventually be discovered, the theory of evolution does not strictly depend on this additional analytical step. If refusal to accept the validity of the theory of evolution is widespread, it raises the question, what should normally make a scientific hypothesis acceptable? A popular misconception is that scientists invariably withhold assent from statements which purport to be scientific until something called 'proof' is provided. It would be more accurate to say that scientists have an inclination to believe any tolerably sound argument of a scientific character until something called 'disproof' is provided. There is, however, an important distinction between a scientific argument and other deductive arguments of equal logical precision, which rests on the contact that scientific arguments make with the real world. The real world does provide the possibility of disproof.

In the case of Darwin's theory of evolution the real world has consistently failed to realise its potential for disproof. Wild organisms *do* vary, a lot of variation *is* inherited, and not all variations confer equal fitness. Natural selection and the evolution of natural populations have been observed directly, even in man. There is now no doubt that natural selection is a mechanism of evolutionary change. Whether it is the exclusive mechanism of evolutionary change is still occasionally debated. If another mechanism for evolutionary change should be discovered, it will have to take its place in a renovated theory of evolution alongside, not instead of, natural selection.

It might be asked why we should award scientific status to claims made by the theory of evolution about the distant past. Surely such claims are not open to disproof, and are therefore in principle unscientific. But is the past really so inscrutable? The fossil record *might* have stopped dead in 4004 BC, but in fact goes back for about three thousand million years. The fossil record *might* have shown the co-existence of all the groups of organisms that we now recognise down to the earliest rocks, but in fact it shows a continual, if occasionally somewhat jerky, development towards present day forms from a very different and more primitive past. Man alone *might* have sprung fully formed into the fossil record, with no evidence of a proto-human ancestry, while in fact there are innumerable forms which look like representatives of our primitive ape-like state. One has to admit that geology does not provide the best evidence for the validity of the modern theory of evolution, but then digging in the rocks is not, by present standards, a very good experiment. As far as the geological evidence goes, however, every opportunity that has been offered to it to provide a refutation of the theory has failed.

The real issue concerning the application of the theory of evolution to unseen events is the same as that raised by any scientific generalisation. It is most satisfying to the scientific cast of mind to suppose that a rule which is confirmed whenever it is tested will apply generally in all cases in which the relevant conditions are met. Since we know nothing about the consequences of the passage of time which might invalidate any of the fundamental conditions for the evolutionary process, and since such evidence as we can lay our hands on is compatible with evolution and with no other hypothesis of comparable scientific stature, we assume that the theory of evolution is generally correct. Indeed it is doubtful whether the theory of evolution would be less widely accepted among scientists if all dead organisms crumbled into dust and we had no fossil record at all.

It is beyond the scope of this book to enquire whether the scientific cast of mind is in any general sense entitled to its confidence in the regularity of events in the external world. In the particular case of the theory of evolution, however, it is difficult to find fault with it. If the theory of evolution is not universally accepted one must assume that it is either due to ignorance of the arguments on which it rests, or else to the unpalatability of its conclusions. More than any other scientific theory, the theory of evolution reaches into hallowed areas of spiritual life. It is no longer easy to look to spiritual authority for a satisfying account of the ultimate issues of human existence: why are we here? why does the world act so uncaringly? what is the sense of the sublime? The Darwinian revolution has been a cruel one in that it has taken away many of the customary sources of consolation. To realise that the physical construction of human bodies and brains is the outcome of processes as comprehensible as those which form the ocean waves may give intellectual satisfaction but it does not necessarily compensate for the loss of divine providence. In the physical continuity between humans, other forms of life and non-living matter, there is no hint of a crack in which humans can find any special dispensation for themselves. Indeed our astonishing intellectual endowments and our vivid sense of self-awareness seem peculiarly well adapted to highlight precisely this deficiency. Because he drew attention to this evident fact, however indirectly, Darwin has been vilified for over a hundred years as the apostle of materialism and a primary source of moral degradation.

Ironically, although it has been hard, perhaps impossible for the orthodox Christian to come to terms with Darwinism, social theorists have had a field day in finding moral lessons in the theory of evolution. When Engels described Marx's materialist theory of history as comparable with Darwin's theory of evolution, one must suppose this was an assessment of relative scientific merit rather than of any specific point of similarity in content or application. Nevertheless the idea of inalienable social or moral progress has, ever since Darwin, looked to the theory of evolution for scientific respectability. But just as Darwin got no inspiration for the development of the theory of evolution from active social progressive movements at the beginning of the nineteenth century, so he saw no reason for the incorporation of the theory of evolution into the new progressive philosophy at the end of his life. As he wrote in 1879, 'What a foolish view seems to prevail in Germany on the connection between Socialism and Evolution through Natural Selection' (L iii. 237).

The extraction of absolute ethical principles from the theory of evolution has been a recurrent theme in post-Darwinian thought. In late Victorian society in England and more especially in America a peculiarly beastly form of social climbing, 'Social Darwinism', was established under Herbert Spencer's slogan 'The survival of the fittest'. The evolutionary law was interpreted to mean victory to the strongest as the necessary condition for progress. As a prescription for social behaviour it justified the worst excesses of capitalist exploitation of labour, 'reasoned savagery' as T. H. Huxley labelled it. Huxley laboured in vain to secure the proper province of Darwinism against the moral fervour of its supporters in the social sciences. Even his own grandson, Julian Huxley, a notable evolutionary biologist of the twentieth century, was unable to resist the appeal of an evolution-based humanist morality. The Darwinian theory of evolution, according to Julian Huxley, has given man

> the assurance that there exists outside of himself a 'power that makes for righteousness'; that he is striving in the same direction as the blind evolutionary forces which were moulding his planet aeons before his appearance; that his task is not to oppose, but to crown the natural order . . .

A similarly factitious exploitation of Darwinian ideas can be found in programmes for godless personal salvation, as in L. Ron Hubbard's 'Dianetics', the philosophical progenitor of the

Scientology cult, which asserts as its 'first law', that 'The dynamic principle of existence is: survive!' We are now faced with the humourless predestinarian onslaught of evolutionary sociobiology, although according to Edward Wilson we are not to be told for another century what remedies should be prescribed for our moral tribulations.

What can one conclude from the incredible heterogeneity of philosophical ideals that have used the Darwinian theory of evolution as a justifying principle? If socialism, *laissez-faire* capitalism, wishy-washy humanism and sociobiological fundamentalism can all find support in Darwin's work, then we must believe either that Darwin was absurdly unclear and inconsistent in his arguments, which he certainly was not, or that the theory of evolution has little if anything to do with ethical prescriptions.

9 Darwin as a scientist: an evaluation

Scientists may be distributed very crudely along an axis which leads from the experimental to the theoretical. The experimental scientist is the one who is constitutionally unable to prevent himself from looking under stones to see what might lie underneath. He is compelled by curiosity and a legitimate conviction that looking under stones is a procedure with a decent likelihood of turning up something as yet unseen. He is exactly that scientist which Newton claimed to be when he said of himself 'I seem to have been only a boy playing on the sea-shore, and diverting myself in now and then finding a smoother pebble or a prettier shell than ordinary, whilst the great ocean of truth lay all undiscovered before me.' (History gives an entirely different picture of Newton, and legend confirms it. There is no suggestion that Newton was struck by the theory of gravitation because he had found that sitting under apple trees was in general a source of unexpected but informative events.) The theoretical scientist is one whose exertions are mental rather than physical, who makes discoveries by taking thought rather than by turning stones. He is exactly that scientist who wrote:

About thirty years ago there was much talk that geologists ought only to observe and not theorize; and I well remember someone saying that at this rate a man might as well go into a gravel-pit and count the pebbles and describe the colours. How odd it is that anyone should not see that all observation must be for or against some view if it is to be of any service. (ML i. 176)

Darwin was a theoretical scientist of quite remarkable power. To the theoretical scientist observations are subservient to explanations. It is the ability of a general statement or argument to subsume innumerable facts which appeals: a fact is only of interest in so far as it is included within or excluded from an argument. The theoretical scientist may look under stones, but he does so not merely because he might find something, but because he already has a clear expectation of finding a particular thing. Indeed to the theoretical scientist the very status of a fact is equivocal: if the observer is not looking 'right' he may not see 'right'. As Darwin noted once, 'I have an old belief that a good observer really means a good theorist' (ML i. 195). His older brother, Erasmus, commenting on the equivocal character of the fossil record as a source of evidence in support of the theory of evolution, put the case for the theoretical scientist even more strongly. 'The *a priori* reasoning is so entirely satisfactory to me that if the facts won't fit in, why so much the worse for the facts . . .' (L ii. 233).

Darwin's first scientific theory, on the origin and distribution of coral reefs, a theory which combined a general understanding of the phenomenon of land subsidence and some characteristics of coral reefs that were extracted from Lyell's *Principles of Geology*, conformed exactly to this pattern.

No other work of mine was begun in so deductive a spirit as this; for the whole theory was thought out on the west coast of S. America before I had seen a true coral reef. I had therefore only to verify and extend my views by a careful examination of living reefs. (A 57)

Darwin had obviously ordered the elements of which the theory was composed into a general idea without the help of a plethora of immediately relevant observations. It was a purely formal argument, scientific because it could be subjected to critical tests, but which initially existed only as a complex mental image, depending upon the faculty which we all possess to some extent of seeing the formal connected consequences of a set of precise but superficially disconnected statements.

Necessarily, scientific theories are not born in a vacuum: a specific issue, fact or collection of facts has to arouse the scientist's attention. A sense of curiosity and of dissatisfaction that at present the issue is unresolved or the facts unexplained leads to a process of speculation whose conclusion is a hypothesis. Finally a further visionary process connects the hypothesis with derivative predictions about the real world which can be looked for directly.

The origins of the theory of evolution by natural selection follow the same pattern, except that one aspect of the theory, namely the idea that the evolutionary hypothesis had explanatory value, was already familiar to Darwin from Lyell's comments on Lamarck. In this case, following the methodology of uniformitarian geology, it was the evolutionary hypothesis itself as well as the fact of adaptation that was unexplained. The sense of curiosity and dissatisfaction is explicit in Darwin's autobiographical reminiscence, 'the subject haunted me' (A 70). On his return from the *Beagle* voyage, Darwin collected 'all facts which bore in any way on the variation of animals and plants under domestication' in the hope that 'some light might perhaps be thrown on the whole subject... I worked on true Baconian principles, and without any theory collected facts on a wholesale scale' (A 71). A few years ago Sir Peter Medawar, a tireless spokesman for the priority of hypothesis over observation in scientific investigation, rebuked Darwin for this 'untrustworthy' description of his actions. In doing so, Medawar seems not to have taken enough account of the existence for a scientist of periods of genuine uncertainty about how an argument should proceed, coupled with the ineffable feeling that there is an argument to be found if only it could be worried at enough. The 'true Baconian principles' of scientific enquiry belong to a pre-Darwinian age when the value of hypothesis in guiding scientific action was hidden behind the plausible fiction that if enough facts could be gathered they would eventually speak for themselves. Darwin certainly recognised the hopelessness of such an enterprise, but seemed anxious to point out that for a brief period after returning from the *Beagle* he really did not know quite what he was looking for.

With the revelation of the struggle for existence the theory of natural selection was virtually complete. At this point a lesser scientist then Darwin would have considered the work adequately done. A formal mechanism for the evolutionary process had been defined and the connection between the explanatory value of the evolutionary hypothesis and the problem of the origin of species was established.

Would this have been a proper time to publish? Darwin's

twenty-year silence about his discovery has provoked endless speculation about the psychopathology of a man who could keep such momentous knowledge under his hat for so long. Although the purely imaginative, i.e. hypothetical, and purely logical, i.e. deductive, phases in scientific argument were well recognised and perfectly respectable in mid-nineteenth century science, to work out an argument fully from its source in the real world to its implications about the real world was certainly seen to be a necessary part of the complete process. Darwin himself was intolerant of purely hypothetical arguments, seeing how easily they can slide from sense to nonsense. A popular formal classification of the early nineteenth century was founded on the hypothesis that real classificatory groups of organisms were ordered in sets of five. When Darwin read that an opponent of the 'quinary' system was proposing a modification to sets of four, he expostulated in his notebooks, 'Anyone may believe anything in such rigmarole about analogies and numbers.' Again, in a thumbnail sketch of the evolutionary philosopher, Herbert Spencer:

> His deductive manner of treating every subject is wholly opposed to my frame of mind. His conclusions never convince me: and over and over again I have said to myself, after reading one of his discussions, 'Here would be a fine subject for half-a-dozen years work' (A 64)

Darwin scanned the real world against the implications of his arguments with extraordinary persistence, using his own observations, and seeking relevant information from any available source, the scientific literature, correspondence and questionnaires. Perhaps Darwin's writing hand was stayed in part by the importance of getting so controversial a theory as evolution by natural selection absolutely right, but the delay, seen in the context of Darwin's overall approach to the scientific enterprise, was completely in character. There has undoubtedly been a shift in emphasis over the last generation about the value of the speculative part of biology in provoking novel observations. It is now fairly commonplace to find hypothetical schemata published in what would have seemed to Darwin to have been an embarrassingly incomplete form. But science languishes for want of imaginative novelty, and there are now many more hands to do the 'work'.

So Darwin delayed his publication until the work had been done, and the place of natural selection in a fully integrated theory of evolution had been secured. I hope that this book gives some idea of the range and versatility of Darwin's exploration of the formal implications of evolution by natural selection. Such brilliant intuitions as the principle of economy, the concept of sexual selection, the identification of the paradox of social behaviour and its solution in selection operating through the family: none of these is at all obvious in the elementary formulation of the theory. Darwin's reputation as a scientist depends not only on the identification of natural selection as the effective process in evolution, but also on the extraordinary completeness of the analysis.

Why is Darwin's name given such overwhelming priority over that of Alfred Russel Wallace in most discussion on the origin of the theory of evolution by natural selection? In point of crispness or clarity there is nothing to choose between Wallace's essay of 1858 and Darwin's brief announcement of his own views, presented with Wallace's at the Linnean Society. Had these been the only productions on evolution by both authors, there would be no reason to rate Darwin more highly than Wallace. Everything that was elegant and imaginative about Darwin's theory of natural selection was equally so in Wallace's. Darwin, of course, had a priority of twenty years in the idea, but none in its publication. Wallace's work was almost certainly entirely independent of Darwin's. The distinction between these two great biologists finally rests on Darwin's relentless pursuit of the idea through all its ramifications to the point where virtually everything except the machinery of inheritance and variation had been clarified.

Wallace's self-denial in the face of Darwin's priority has a magnificence for which he should be remembered as vividly as for his co-discovery of the mechanism of evolution. As any scientist knows, a hypothesis is as precious to its discoverer as a child to its parents. Yet Wallace found the strength of character to entitle two of his books *Natural Selection* (Darwin's phrase) and *Darwinism*, and to write in the Preface to the former what will also serve as a coda to this book.

> I have felt all my life, and I still feel, the most sincere satisfaction that Mr. Darwin had been at work long before me, and that it was not left for me to attempt to write 'The Origin of Species'. I have long since measured my own strength, and know well that it would be quite unequal to that task. Far abler men than myself may confess, that they have not that untiring patience in accumulating, and that wonderful skill in using, large masses of facts of the most varied kind, – that wide and accurate physiological knowledge, – that acuteness in devising and skill in carrying out experiments, – and that admirable style of composition, at once clear, persuasive and judicial, – qualities, which in their harmonious combination, mark out Mr. Darwin as the man, perhaps of all men now living, best fitted for the great work he has undertaken and accomplished.

Further Reading

It would be convenient to be able to refer to a uniform edition of Darwin's books, but none exists. Only *The Origin of Species* (see Abbreviations, and 6th Edition available in Penguin), *The Voyage of the Beagle* (Penguin), *The Structure and Distribution of Coral Reefs* (University of California Press, 1977), and *The Expression of the Emotions* (University of Chicago Press, 1965) are likely to be easily available. A reprint of *The Descent of Man* is planned by Princeton University Press. Nearly all of Darwin's Notebooks are now in print (see Abbreviations), and his scientific papers published in learned journals have now been collected and edited by Paul H. Barrett (University of Chicago Press, 1977). For Darwin's other works one must go to libraries or second-hand bookshops. He has had the privilege of an excellent bibliography by R. B. Freeman (Wm. Dawson & Sons, Ltd., Folkestone, 1977).

For general studies, Sir Gavin de Beer's *Charles Darwin* (1963, reprinted by Greenwood Press, 1976) is an excellent scientific biography, while Michael Ghiselin's *The Triumph of the Darwinian Method* (University of California Press, 1969) is the most enjoyable of all the books which attempt a synthetic appraisal of Darwin's entire scientific work.

Stephen Jay Gould's two books of essays, *Ever Since Darwin* (Burnett Books, 1978) and *The Panda's Thumb* (Norton, 1980) are clever, informative and often very funny. I also recommend the first part of his major book, *Ontogeny and Phylogeny* (Belknap, Harvard University Press, 1977), whose title consists of two technical terms that I have managed to avoid, largely by hoping that my readers will develop their understanding of the conceptual relationship between embryonic and evolutionary development under Professor Gould's instruction.

Three recent products of the Darwin industry deal with some wider contexts for Darwinism in the nineteenth century: Neil C. Gillespie's *Charles Darwin and the Problem of Creation* (University of Chicago Press, 1979); James R. Moore's *The Post-*

Darwinian Controversies (Cambridge University Press, 1979) and Michael Ruse's *The Darwinian Revolution* (University of Chicago Press, 1979).

Pre-Darwinian rationalisations of the natural world were dealt with beautifully by Arthur O. Lovejoy in *The Great Chain of Being* (1936), a classic which is still available in paperback (Harvard University Press). Pre-Darwinian evolutionary speculations and many of the themes covered in chapter 2 are developed in detail in *Forerunners of Darwin 1745–1859* (The Johns Hopkins Press, 1968), edited by Bentley Glass, Owsei Temkin and William L. Strauss, Jr.

For modern evolution, Richard Dawkins's *The Selfish Gene* (Paladin, 1978) is a kind of parable dealing with the evolutionary process entirely from the gene's point of view. Dawkins's main interest is in animal behaviour and he is imaginative and humane on the difficult problems of evolution of social relations. Colin Patterson's *Evolution* (British Museum (Natural History, 1978) is a good, simple, well-illustrated account of the present state of evolutionary biology. More difficult, but exceptionally good, is John Maynard Smith's *The Theory of Evolution* (Penguin, 1978).

Index

Past Masters

HUME A. J. Ayer

A. J. Ayer begins his study of Hume's philosophy with a general account of Hume's life and works, and then discusses his philosophical aims and methods, his theories of perception and self-identity, his analysis of causation, and his treatment of morals, politics and religion. He argues that Hume's discovery of the basis of causality and his demolition of natural theology were his greatest philosophical achievements.

JESUS Humphrey Carpenter

Humphrey Carpenter writes about Jesus from the standpoint of a historian coming fresh to the subject without religious preconceptions. He examines the reliability of the Gospels, the originality of Jesus's teaching, and Jesus's view of himself. His highly readable book achieves a remarkable degree of objectivity about a subject which is deeply embedded in Western culture.

KANT Roger Scruton

Immanuel Kant is perhaps the greatest, and certainly the most original, of modern philosophers. He is also among the most difficult to understand. In tackling this exceptionally complex subject, Roger Scruton carefully explains Kant's attempt to construct a philosophy that was neither rationalist nor empiricist, and shows why his great work, the *Critique of Pure Reason*, has proved enduringly influential.